WITHDRAWN

THE NATURE OF PLAY

The Nature of Play

GREAT APES AND HUMANS

Edited by
ANTHONY D. PELLEGRINI
PETER K. SMITH

THE GUILFORD PRESS
New York London

A Division of Guilford Publications, Inc.
72 Spring Street, New York, NY 10012
www.guilford.com

Printed in the United States of America

This book is printed on acid-free paper.

Last digit is print number: 9 8 7 6 5 4 3 2 1

Library of Congress Cataloging-in-Publication Data

The nature of play : great apes and humans / edited by Anthony D. Pellegrini, Peter K. Smith.
p. cm.
Includes bibliographical references and index.
ISBN 1-59385-117-0 (hardcover)
1. Play—Psychological aspects. 2. Play behavior in animals. 3. Psychology, Comparative. I. Pellegrini, Anthony D. II. Smith, Peter K.
BF717.N38 2004
156.′5—dc22

2004020577

About the Editors

Anthony D. Pellegrini, PhD, is Professor of Psychological Foundations of Education in the Department of Educational Psychology at the University of Minnesota, Twin Cities. His primary interest is in the development of play and dominance. He also has research interests in methodological issues in the general area of human development, with specific interest in direct observations. His research has been funded by the National Institutes of Health, the Spencer Foundation, and the W. T. Grant Foundation. Dr. Pellegrini is a Fellow of the American Psychological Association and has been awarded a Fellowship from the British Psychological Society.

Peter K. Smith, PhD, is Professor of Psychology and Head of the Unit for School and Family Studies at Goldsmiths College, University of London. His research interests are in social development, play, bullying in school, and evolutionary theory. Dr. Smith is coauthor of *Understanding Children's Development* (4th ed.; Blackwell, 2003) and coeditor of *The Nature of School Bullying* (Routledge, 1999) and the *Blackwell Handbook of Childhood Social Development* (Blackwell, 2002). He has written widely on children's play, particularly on pretend play training and rough-and-tumble play. Dr. Smith is also a Fellow of the British Psychological Society.

Contributors

Patrick Bateson, PhD, Department of Zoology, Kings College, University of Cambridge, Cambridge, United Kingdom

John Bock, PhD, Department of Anthropology, California State University, Fullerton, California

Vera Silvia Raad Bussab, PhD, Department of Experimental Psychology, University of São Paolo, São Paolo, Brazil

Douglas P. Fry, PhD, Department of Social Sciences, Abo Akademi University, Vaasa, Finland, and Bureau of Applied Research in Anthropology, University of Arizona, Tucson, Arizona

Juan-Carlos Gómez, PhD, School of Psychology, University of St. Andrews, St. Andrews, United Kingdom

Yumi Gosso, MS, Department of Experimental Psychology, University of São Paolo, São Paolo, Brazil

Kathy Gustafson, MS, Department of Educational Psychology, University of Minnesota, Minneapolis, Minnesota

Kerrie P. Lewis, PhD, Department of Anthropology, Washington University, St. Louis, Missouri

Beatriz Martín-Andrade, MS, School of Psychology, University of St. Andrews, St. Andrews, United Kingdom

William C. McGrew, PhD, Departments of Anthropology and Sociology and Gerontology, Miami University, Oxford, Ohio

Maria de Lima Salum e Morais, MS, São Paolo State Health Department, São Paolo, Brazil

Emma Otta, PhD, Department of Experimental Psychology, University of São Paolo, Sao Paolo, Brazil

Anthony D. Pellegrini, PhD, Department of Educational Psychology, University of Minnesota, Minneapolis, Minnesota

Jacklyn K. Ramsey, BA, Department of Anthropology, Miami University, Oxford, Ohio

Fernando José Leite Ribeiro, PhD, Department of Experimental Psychology, University of São Paolo, São Paolo, Brazil

Peter K. Smith, PhD, Department of Psychology, Goldsmiths College, University of London, London, United Kingdom

Contents

PART IV
FANTASY

PART V
HUNTER-GATHERERS AND PASTORAL PEOPLES

PART VI
CONCLUSION

THE NATURE OF PLAY

PART I

BACKGROUND AND THEORY

CHAPTER 1

Play in Great Apes and Humans

ANTHONY D. PELLEGRINI AND PETER K. SMITH

Over the past few years a number of scholars has examined human development in its larger evolutionary context (e.g., Archer, 1992; Bateson & Martin, 1999; Bjorklund & Pellegrini, 2002; Campbell, 2002; Laland & Brown, 2002). This current interest in the role of evolutionary theory in human development is, in some ways, an interesting return to the origins of developmental psychology as a separate branch of the field (Bjorklund & Pellegrini, 2002; Pellegrini, 2004).

HISTORICAL CONTINUITY

G. Stanley Hall (1904, 1916), called by many (e.g., Cairns, 1983) the father of child and developmental psychology in the United States, used Darwin's theory of evolution by natural selection to formulate a theory of human development. Hall is remembered for his emphasis of the stage-like progression of human development. For Hall, ontogenetic stages reenacted the phylogenetic history of the species. In this vein he is also remembered for the phrase "ontogeny recapitulates phylogeny," meaning that development within a species repeats the evolutionary history of the species. For example, the stages of child development were said to repeat the history of *Homo sapiens.* Some of Hall's famous examples include boys' tree climbing as a recapitulation of our primate past, and boys' play fighting (then it took the form of "cowboys and Indians") as a reenactment of our hunter-gatherer past.

Of course, the accuracy of the notion that ontogeny repeats phylogeny is questionable. Specifically, and without going into detail (see Hinde, 1983 for extended discussion), it is more likely that ontogenetic development influences phylogeny, as Bateson (Chapter 2, this volume) suggests. The course of individual development, for example, the age of sexual maturity, influences the development of the species.

We should not, however, be too hasty and chuckle too visibly at the apparent foolishness of Hall's claim. Most theories are destined not to be supported. Newton's theory was displaced by Einstein's (Clark, 1971), but this does not discredit the historical importance of Newton for the physics of his day.

Probably the most enduring aspect of Hall's theory, and his importance for this volume, was his emphasis on the stages of human development and the corresponding importance he placed on the role of play during childhood. For Hall play was integral to childhood; it was a period during which children should be allowed unfettered time to explore and play. He warned against the use of "unnatural" and "artificial" educational and pedagogical regimens, stating that they interfered with the natural unfolding of children's developmental processes (Cairns, 1983, p. 52). Play, for Hall, meant children choosing an activity and enacting it on their own terms. In this regard play during childhood was stressed as a way in which children expressed their phylogenetic history.

Perhaps more interesting was Hall's view that play was *not* something that prepared individuals for adulthood. Unlike most of his contemporaries, such as Groos (1898, 1901) and later Piaget (1962) and Vygotsky (1967), Hall considered play important to childhood, not as practice for adulthood. Indeed, this view is current among many developmental psychologists (e.g., Bjorklund & Pellegrini, 2002; Pellegrini & Smith, 1998) and evolutionary biologists (Bateson, 1981; Chapter 2, this volume)!

Hall's work emphasized the notions of the juvenile period as a distinctive "stage" of human development. Correspondingly, play was seen as the way in which juveniles come to know their world. Through play, juveniles interact with their physical and social worlds and "construct" their mental worlds. The role of play in development has since been discussed by child psychologists (Pellegrini, 2003; Power, 2000; Smith, 1982), comparative psychologists (Harlow, 1962; Pellis & Pellis, 1998), and ethologists (Bateson, 1981; Bateson & Martin, 1999; Burghardt, 2004). Thus, it is a natural candidate for cross-species comparison.

Play manifests itself behaviorally in similar ways across many mammalian species (for a masterful synthesis, see Fagen, 1981). For example, in the case of rough-and-tumble play (R&T), or play fighting, the behaviors and social roles differentiating it from real fighting show some similarity in human juveniles (Blurton Jones, 1972; Fry, Chapter 4, this volume; Humphreys & Smith, 1987; Pellegrini, 2003), rhesus monkeys (Harlow, 1962; Suomi, 2002), and laboratory rats (Pellis & Pellis, 1998). In terms of behavior, R&T is realized through soft, not hard, hits, and the bigger and stronger individuals often engage in handicapping roles so as to maintain playful interaction.

NOT A "MAIN-EFFECT" VIEW OF EVOLUTIONARY HISTORY ON HUMAN BEHAVIOR: A TRANSACTIONAL VIEW

Although we believe that evolutionary history has an impact on human development, we must further clarify the moderating role of context. This discussion is especially important given recent misunderstanding of the evolutionary developmental psychology approach (Lickliter & Honeycutt, 2003). We do not take a deterministic view of the impact of phylogenetic history and genes on human behavior, or what Archer (Archer & Lloyd, 2002) calls a "main effect model," even if continuity across common ancestry is found in dimensions of social behavior and organization (Bjorklund & Pellegrini, 2002). Our orientation leads to the position that genes, environments, and behavior dynamically influence each other (Archer & Lloyd, 2002; Bateson & Martin, 1999; Bjorklund & Pellegrini, 2002; Gottlieb, 1998; Pellegrini & Archer, in press; Stamps, 2003).

By way of analogy, this orientation is akin to adding another, more distal, level to Bronfenbrenner's (1979; Bronfenbrenner & Ceci, 1994) model of human development. For example, although this model has a layer for socioeconomic status (SES), this does not translate into a main effect for SES. Instead, SES is seen as influenced by and influencing adjacent layers of context. This dynamic relation is illustrated in work on "resilient" children (Masten & Coatsworth, 1998). Aspects of children's temperament and family moderate the effects of deleterious environments associated with low SES. By including an evolutionary layer, we are assuming that evolution by natural selection (rather than other explanations) influences, at a distal level, our current status as humans. This status, however, is a dynamic one that interacts with more proximal aspects of children' lives.

Specifically, the environment in which an individual develops, starting with conception, influences the ways in which evolutionary history is expressed. Following Bateson and Martin's (1999) juke-box metaphor, individuals have a genetic endowment that can be realized through a wide variety of options, but the specific developmental pathway taken by an individual is influenced by the perinatal environment (i.e., from conception through infancy) of the developing organism. Thus, a number of developmental pathways are possible, but which one is selected is determined by the environment in which the organism develops (Archer, 1992; Caro & Bateson, 1986).

A relevant aspect of the environment is the effect of nutrition on sexual dimorphism in size, which, in turn, affects physically vigorous behavior, sex segregation, and play behavior. Specifically, the nutritional history of human mothers impacts the physical size of the offspring, especially that of males (Bateson & Martin, 1999). Males' larger size, relative to females' (i.e., sexual dimorphism) may be one of the factors responsible for sex differences in social play, wherein males engage in R&T more than females (Pellegrini, 2004b; in press).

Consistent with this view, the availability of resources affects human mating systems and sexual dimorphism. The interactive nature of this system is illustrated by the fact that height is highly heritable, yet this relation is influenced by the fetus's perinatal environment; pregnant mothers experiencing nutritional stress will have relatively smaller offspring; this is especially the case for male offspring (Bateson & Martin, 1999).

Alexander, Hoogland, Howard, Noonan, and Sherman (1979) used the Human Relations Area File to examine ecologically abundant and stressed societies. They also partitioned human societies into ecologically imposed monogamy (e.g., Lapps of Norway, Cooper and Labrador Eskimo), polygyny (e.g., Bedouin Arabs and Khmer), and culturally imposed monogamy (most Western societies). They found that the ecologically imposed groups lived in less abundant ecologies and were significantly less dimorphic than the other two, but there were no differences between the polygynous and culturally imposed monogamous groups. They argued that in ecologically imposed monogamous cultures, efforts of both parents are needed to protect and provision offspring. From this view, sexual dimorphism results from the confluence of ecological conditions and mating systems. Dimorphism is, in turn, hypothesized to be an antecedent condition for different energetic demands of males and females as well as an antecedent to both sex segregation and sex differences in play (Pellegrini, 2004b; in press).

Differences in body size are antecedents to differences in physical activity and associated with competitiveness, aggressiveness, and play, such that males and females view themselves differently very early in development and then segregate (see Pellegrini, in press, for a fuller discussion). Later, these differences are translated into different social roles, differences in R&T, and agonistic behaviors associated with different energetic demands.

In short, the developmental option taken is related to the environment in which the individual develops. Given the vastly different environments into which individuals with the same genetic history are born, a single "genetic program" would not be equally effective across these different niches. This model is displayed in Figure 1.1.

CHOOSING COMPARISONS: SPECIES AND HUMAN SOCIAL ORGANIZATION

It is common practice among evolutionary biologists to demonstrate the extent to which specific behaviors, such as play, may have resulted from similar selection pressures, and to compare the development, function, and causes of these behaviors across different species (Tinbergen, 1963). In choosing species comparisons, however, we are faced with a number of

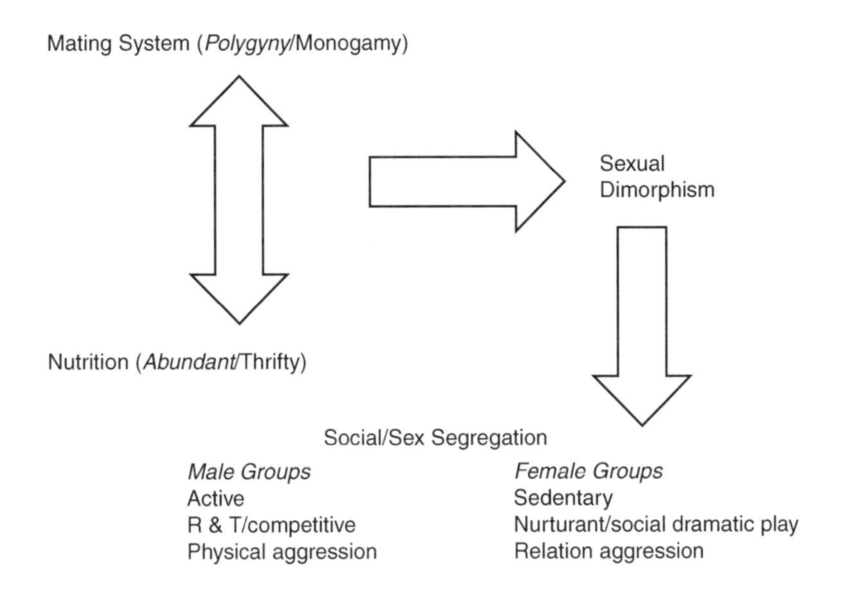

FIGURE 1.1. A sexual selection model for sex differences in R&T and aggression.

choices. At one, and perhaps the least satisfactory, level we can choose species, ad hoc, as a basis for comparison; for example, by comparing human behavior in one case to birds and in other cases to monkeys, and in yet other cases to apes. Such comparisons may (correctly) raise the specter that we are merely constructing "just-so" stories. Comparing humans with the great apes is more satisfactory: the chimpanzees (*Pan troglodytes* and *Pan schweinfurthii*), bonobos (*Pan paniscus*), gorillas (*Gorilla gorilla*), and orangutans (*Pongo pygmaeus*). The chimpanzees and bonobos are our closest phylogenetic relatives (Wrangham, Chapman, Clark-Arcadi, & Isabirye-Basuta, 1996), as indicated by similarities in DNA evidence, and have similar dispersal patterns to humans (with males staying and females often transferring to another group). They have been posited as especially informative comparator species (McGrew, 1981; deWaal & Lanting, 1997).

Female dispersal, male residency, and polygyny are especially important when we examine social and locomotor play. For example, when males, not females, stay in the natal groups (females migrate), their social behavior is more gregarious and their social groups become organized hierarchically, in terms of dominance. Both females and males segregate into same-sex groups to learn associated sex-specific roles. Specifically, the social skills learned in segregated groups should be related to males' and females' adult roles. This correspondence appears to be the case in chimpanzees; for example, males learn dominance and predation roles, such as patrolling (Mitani, Merriweather, & Zhang, 2000), and females learn to handle infants (Pusey, 1990). The case of male chimpanzees segregating is especially robust. The complexity of the skills associated with adult life may necessitate extended practice, and social play has been proffered as a context in which these skills have evolved (Alexander, 1989; Wrangham, 1999).

The juvenile and adolescent periods provide the social context to learn and practice adult roles. For males, skills related to dominance, such as detecting weaknesses and coordinating skilled movements, may need extensive practice, perhaps through play with conspecifics, given their variety and complexity (Alexander, 1989). That dominance skills are especially important to males' functioning (Wrangham, 1999) points to the development of these activities taking place in a social context. Indeed, human cross-cultural studies show that male juveniles are socialized into more competitive and aggressive roles, especially in polygynous societies where roles are not stratified (Low, 1989). In female groups, too, roles are learned and practiced, and physically aggressive males are avoided. Females interact with other females in dyads, and in these smaller groups use safer, indirect aggression to form alliances and coalitions with other females (Campbell, 2002).

THE ORDER OF THINGS

This volume is organized to reflect the ways in which ethologists and developmental psychologists think about play. Traditionally, ethologists created categories in terms of object play, social play, and locomotor or exercise play (Martin & Caro, 1985), with little or no attention given to fantasy or pretend play. Developmental psychologists, on the other hand, tended to study fantasy play more than either social or object play and paid little or no attention to locomotor play (see Pellegrini & Smith, 1998; Power, 2000). When locomotor play has been studied by developmentalists, it has typically occurred only in the context of social play, especially in the form of R&T (Pellegrini, 2002; Power, 2000; Smith, 1997).

In many ways these differences in categorizing play reflect the multidimensional nature of the construct. To accommodate these different ways of categorizing play and recognizing the considerable overlap among categories, we parse play in terms of social, object, and fantasy, for both humans (in industrialized, pastoral, and foraging societies) and great apes.

We also chose to include chapters that address the play of both foraging and pastoral societies. This decision was based on a basic assumption of "evolutionary psychology" (questioned by many; see Laland & Brown, 2002) that the behavior of foragers (or hunter-gatherers) may represent the "environment of evolutionary adaptedness" (EEA)—that is, the environment in which our current genetic makeup developed. Inclusion of research on the play of foragers is valuable, even if we remain skeptical of the idea of the EEA, because we simply do not know much about these groups. Archiving their behavior is especially important in light of the rapidity with which these groups are disappearing. Additionally, when added to comparisons of studies with pastoral and contemporary societies, we gain a broader understanding of the contexts in which human behavior developed.

This book is organized into sections. Bateson sets the scene in Chapter 2 with a theoretical overview of the role of play. Subsequent sections pair discussions of various types of play in great apes and in humans: social play (Chapter 3, Lewis; Chapter 4, Fry), object play (Chapter 5, Ramsey & McGrew; Chapter 6, Pellegrini & Gustafson), and fantasy play (Chapter 7, Gómez & Martín-Andrade; Chapter 8, Smith). Much of the evidence on human play cited in these chapters comes from modern urban societies, so the next section focuses very specifically on play in human hunter-gatherer communities (Chapter 9, Gosso and colleagues) and foraging societies (Chapter 10, Bock). In a concluding chapter we reflect on the various contributions and on the continuities and discontinuities in the play of great apes and different human societies.

REFERENCES

Alexander, R. D. (1989). Evolution of the human psyche. In P. Mellers & C. Stringer (Eds.), *The human revolution: Behavioral and biological perspectives on the origins of modern humans* (pp. 455–513). Princeton, NJ: Princeton University Press.

Alexander, R. D., Hoogland, J. L., Howard, R. D., Noonan, K. M., & Sherman, P. W. (1979). Sexual dimorphisms and breeding systems in pinnipeds, ungulates, primates, and humans. In N. A. Chagnon & W. Irons (Eds.), *Evolutionary biology and human social behavior* (pp. 402–435). North Scituate, MA: Duxbury Press.

Archer, J. (1992). *Ethology and human development.* Hemel Hempstead, UK: Harvester Wheatsheaf.

Archer, J., & Lloyd, B. (2002). *Sex and gender* (2nd ed.). London: Cambridge University Press.

Bateson, P. P. G. (1981). Discontinuities in development and changes in the organization of play in cats. In K. Immelmann, G. Barlow, L. Petrinovich, & M. Main (Eds.), *Behavioral development* (pp. 281–295). New York: Cambridge University Press.

Bateson, P. P. G., & Martin, P. (1999). *Design for a life: How behaviour develops.* London: Cape.

Bjorklund, D. F., & Pellegrini, A. D. (2002). *Evolutionary developmental psychology.* Washington, DC: American Psychological Association.

Blurton Jones, N. (1972). Categories of child–child interaction. In N. Blurton Jones (Ed.), *Ethological studies of child behaviour* (pp. 97–129). London: Cambridge University Press.

Bronfenbrenner, U. (1979). *The ecology of human development.* Cambridge, MA: Harvard University Press.

Bronfenbrenner, U., & Ceci, S. J. (1994). Nature–nurture reconceptualized in developmental perspective: A bioecological model. *Psychological Review, 101*, 568–586.

Burghardt, G. (2004). *The genesis of animal play: Testing the limits.* Cambridge, MA: MIT Press.

Cairns, R. (1983). The emergence of developmental psychology. In W. Kessen (Ed.), *Handbook of child psychology* (Vol. 1, pp. 41–102). New York: Wiley.

Campbell, A. (2002). *A mind of her own: The evolutionary psychology of women.* Oxford, UK: Oxford University Press.

Caro, T. M., & Bateson, P. (1986). Ontogeny and organization of alternative tactics. *Animal Behaviour, 34*, 1483–1499.

Clark, R. W. (1971). *Einstein: The life and times.* New York: Avon Books.

deWaal, F., & Lanting, F. (1997). *Bonobo: The forgotten ape.* Berkeley: University of California Press.

Fagen, R. M. (1981). *Animal play behavior.* Oxford, UK: Oxford University Press.

Gottlieb, G. (1998). Normally occurring environmental and behavioral influences on gene activity: From central dogma to probabilistic epigenesis. *Psychological Review, 105*, 792–802.

Groos, K. (1898). *The play of animals.* New York: Appleton.

Groos, K. (1901). *The play of man.* London: Heinemann.

Hall, G. S. (1904). *Adolescence: Its psychology and its relation to physiology, anthropology, sociology, sex, crime, religion, and education* (Vols. 1 & 2). New York: Appleton.

Hall, G. S. (1916). *Adolescence.* New York: Appleton.

Harlow, H. (1962). The heterosexual affection system in monkeys. *American Psychologist, 17,* 1–9.

Hinde, R. A. (1983). Ethology and child development. In J. J. Campos & M. H. Haith (Eds.), *Handbook of child psychology: Vol. II. Infancy and developmental psychobiology* (pp. 27–94). New York: Wiley.

Humphreys, A., & Smith, P. K. (1987). Rough-and-tumble play, friendship and dominance in school children: Evidence for continuity and change with age. *Child Development, 58,* 201–212.

Laland, K. N., & Brown, G. R. (2002). *Sense and nonsense: Evolutionary perspectives on human behaviour.* Oxford, UK: Oxford University Press.

Lickliter, R., & Honeycutt, H. (2003). Developmental dynamics: Toward a biologically plausible evolutionary psychology. *Psychological Bulletin, 129,* 819–835.

Low, B. S. (1989). Cross-cultural patterns in the training of children: An evolutionary perspective. *Journal of Comparative Psychology, 103,* 311–319.

Martin, P., & Caro, T. (1985). On the function of play and its role in behavioral development. In J. Rosenblatt, C. Beer, M-C. Bushnel, & P. Slater (Eds.), *Advances in the study of behavior* (Vol. 15, pp. 59–103). New York: Academic Press.

Masten, A. S., & Coatsworth, J. D. (1998). The development of competence in favorable and unfavorable environments. *American Psychologist, 53,* 205–220.

McGrew, W. C. (1981). The female chimpanzee as a female evolutionary prototype. In F. Dahlberg (Ed.), *Woman the gatherer* (pp. 35–73). New Haven, CT: Yale University Press.

Mitani, J. C., Merriwether, A., & Zhang, C. (2000). Male affiliation, cooperation and kinship in wild chimpanzees. *Animal Behaviour, 59,* 885–893.

Pellegrini, A. D. (2002). Rough-and-tumble play from childhood through adolescence: Development and possible functions. In P. K. Smith & C. H. Hart (Eds.), *Blackwell handbook of childhood social development* (pp. 438–454). Oxford, UK: Blackwell.

Pellegrini, A. D. (2003). Perceptions and possible functions of play and real fighting in early adolescence. *Child Development, 74,* 1459–1470.

Pellegrini, A. D. (2004a). *Observing children in the natural worlds: A methodological primer* (2nd ed.). Mahwah, NJ: Erlbaum.

Pellegrini, A. D. (2004b). Sexual segregation in childhood: A review of evidence for two hypotheses. *Animal Behaviour, 68,* 435–443.

Pellegrini, A. D. (in press). Sexual segregation in humans. In K. Ruckstuhl & P. Neuhaus (Eds.), *Sexual segregation in vertebrates.* Cambridge, UK: Cambridge University Press.

Pellegrini, A. D., & Archer, J. (in press). Sex differences in competitive and aggressive behavior: A view from sexual selection theory. In B. J. Ellis & D. F. Bjorklund (Eds.), *Origins of the social mind: Evolutionary psychology and child development.* New York: Guilford Press.

Pellegrini, A. D. & Smith, P. K. (1998). Physical activity play: The nature and function of a neglected aspect of play. *Child Development, 69*, 577–598.

Pellis, S. M., & Pellis, V. V. (1998). The structure–function interface in the analysis of play fighting. In M. Bekoff & J.A. Byers (Eds.), *Animal play: Evolutionary, comparative, and ecological perspectives* (pp. 115–140). New York: Cambridge University Press.

Piaget, J. (1962). *Play, dreams, and imitation in childhood* (C. Gattengno & F. M. Hodgson, Trans.). New York: Norton. (Original work published 1951)

Power, T. G. (2000). *Play and exploration in children and animals.* Mahwah, NJ: Erlbaum.

Pusey, A. E. (1990). Behavioural changes at adolescence in chimpanzees. *Behaviour, 115*, 203–246.

Smith, P. K. (1982). Does play matter? Functional and evolutionary aspects of animal and human play. *Behavioral and Brain Sciences, 5*, 139–184.

Smith, P. K. (1997). Play fighting and real fighting: Perspectives on their relationship. In A. Schmitt, K. Atswanger, K. Grammer, & K. Schafer (Eds.), *New aspects of human ethology* (pp. 47–64). New York: Plenum Press.

Stamps, J. (2003). Behavioural processes affecting development: Tinbergen's fourth question comes to age. *Animal Behaviour, 66*, 1–13.

Suomi, S. J. (2002, August). *Genetic and environmental contributions to deficits in rough-and-tumble play in juvenile rhesus monkey males.* Paper presented at the biennial meeting of the International Society for the Study of Behavioral Development, Ottawa, Ontario, Canada.

Tinbergen, N. (1963). [On the aims and methods of ethology.] *Zeitschirift für Tierpsychologie, 20*, 410–413.

Vygotsky, L. (1967). Play and its role in the mental development of the child. *Soviet Psychology, 12*, 62–76.

Wrangham, R. W. (1999). Evolution of coalitionary killing. *Yearbook of Physical Anthropology, 42*, 1–30.

Wrangham, R. W., Chapman, C. A., Clark-Arcadi, A. P., & Isabirye-Basuta, G. (1996). Social ecology of Kanyawara chimpanzees: Implications for understanding the costs of great ape groups. In W. C. McGrew, L. F. Marchant, & N. Toshisada (Eds.), *Great ape societies* (pp. 45–57). Cambridge, UK: Cambridge University Press.

CHAPTER 2

The Role of Play in the Evolution of Great Apes and Humans

PATRICK BATESON

Individuals are active agents in their own development, seeking out and acquiring experiences that will shape their future behavior. Many have argued that one of the ways in which young animals and humans do this is by playing (Bateson & Martin, 1999). But what is play? It is a question that continues to tease—even though many fine minds have grappled with the issue for more than a quarter of a century. The question of definition is not trivial, and providing hard evidence to support any of the innumerable proposals for the biological function of play has proved elusive. The role of play in evolution, the topic of this chapter, is, by its very nature, speculative. Nonetheless, I address all three thorny issues: definition, biological function, and role in evolution.

Establishing categories of behavior that command agreement and provide the basis for measurement is a crucial part of behavioral biology. In the case of play the testing issue of definition is often shrugged off because, it is claimed, we all recognize play when we see an individual doing it. However, this is to confuse recognition with agreement. When a young chimpanzee plays, observers readily agree about the occurrences of a variety of different components of its activities; their quantified measurements correlate strongly with each other. But when a fish plays, if it does, how many will state confidently what it is doing? What is the basis for the classification of behavior as play or non-play? I suspect that a major element here is the extent to which the animal's behavior corresponds to what humans do when playing—a not uncommon problem in the classification of many categories of animal behavior. Humans undoubtedly project their notions of playfulness into the seemingly purposeless behavior of other animals. Once classified, it is easy to provide an ostensive definition, pointing to the behavior patterns in question and saying: "That's what I mean by 'play.'" Such definitions are typically accompanied by clear descriptions, drawings, and videos.

All that stated, play, as agreed on by independent observers, is typically something that children and young animals do. Adults play too, of course, but generally have less time for it and less inclination. Most human adults forget what it was like to spend the whole day on the beach with nothing but a bucket and spade and perhaps a friend, doing something that seems—to adults—to be entirely pointless.

Human play comes in many different forms: solitary, imaginary, symbolic, verbal, social, constructional, rough-and-tumble, manipulative, and so forth. The play of a 4-year-old boy wrestling with another 4-year-old is descriptively quite different from that of, say, a solitary 10-year-old staring into space while indulging in a private fantasy about being a pop star or a doctor.

In all its manifestations play is characterised by its apparent lack of serious purpose or immediate goal. Play is the antithesis of adult "work," in which the behavior has an obvious, and usually short-term, goal. Activities are more likely to be perceived as play than work by humans engaged in them, if they are entered into voluntarily. In one experiment, volunteers were given a problem-solving game to perform. Some were paid to perform the game, and some were not. Those who were paid spent less of their free time performing than those for whom the only motivation was the intrinsic pleasure of the game itself (Deci & Ryan, 1980). Motivation to play springs from within, and the readiness to perform tasks may, paradoxically, be reduced by external rewards. A person's eagerness to play increases if the task is freely chosen and the performer discovers that his or her skill at some challenging task improves with practice. Success in performing the task leads to greater enjoyment and hence greater motivation to carry on with the activity. Such is the mainspring of many sports and hobbies.

Play has been described in a wide variety of young animals—most mammals and some bird species, such as parrots and ravens—and is probably much more widespread in other taxonomic groups such as fish than is commonly believed (Fagen, 1981). It may occupy a substantial proportion of their time as well (Martin & Caro, 1985). At the stage in life when a bird or mammal does it most, play can account for around 10% of its time—not as much, perhaps, as a child's, but still a lot.

Playful behavior often resembles "real" behavior but lacks its normal biological consequences: The young animal plays at fighting or catching imaginary prey, but it is usually obvious that the animal *is* playing rather than merely being incompetent. In social play, for instance, the roles of the play partners are frequently reversed and sexual components are often incorporated long before the animal is sexually mature. During play involving running, jumping, and other rapid movements, the movement patterns tend to be exaggerated in form, jumbled in sequence, and often repeated. In some species, specific social signals are used to denote that what follows is playful rather than serious. Dogs, for example, signal their readiness to play by dropping down on their forelegs and wagging their tails, and chimpanzees have a special "play face" that precedes a bout of social play. In the solitary manipulations of object play, the prey-catching or food-getting repertoires of adults are frequently used long before they bring in any real food. Play is also exquisitely sensitive to prevailing conditions and is usually the first nonessential activity to go when all is not well. Its presence or absence is a sensitive barometer of the individual animal's psychological and physical well-being. For instance, young vervet monkeys in East Africa do not play in dry years, when food is scarce (Lee, 1984). Play happens only when basic short-term needs have been satisfied, and the animal is relaxed. It is therefore the first activity to disappear if the animal is stressed, anxious, hungry, or ill.

BIOLOGICAL FUNCTIONS OF PLAY

Nobody doubts the motivation for play, but what is it all *for*? Play has real biological costs. Animals expend more energy and expose themselves to greater risks of injury and predation when they are playing than when they are resting. Play makes them more conspicuous and less vigilant. For example, young Southern fur seals are much more likely to be killed by sea lions when they are playing than at other times (Harcourt, 1991). The costs of play must presumably be outweighed by its benefits, otherwise animals that played would be at a disadvantage compared with those that did not, and play behavior would not have evolved. What are the biological functions of play?

The belief that children's play has a serious purpose—that of acquiring the skills and experience needed in adulthood—has been a central feature in thinking about the nature of play behavior throughout the history of the field. It is commonly thought that play builds adult behavior (Smith, 1982). However, likening the development of behavior to the assembly of buildings is only partly successful as an image, because half-assembled animals, unlike half-assembled buildings, have to survive and find for themselves the materials they need for further construction. Nevertheless, one building metaphor—the use of scaffolding—is helpful in understanding the nature of development. Scaffolding is required for the building process but is usually removed once the job is complete. Play is, in the commonest view of its function, developmental scaffolding. Once this job is done, it largely falls away.

Active engagement with the environment has great benefits, it is argued, because the world is examined from different angles—and the world rarely looks the same from different angles. Such engagement helps to construct a working knowledge of the environment: the recognition of objects, understanding what leads to what, discovering that things are found when stones are turned over and the world is rearranged, learning what can and cannot be done with others. All these discoveries are real benefits for the animal.

The precise nature of the benefits of play remains, nonetheless, a matter of dispute, with little hard evidence to distinguish between the possibilities. The list of putative benefits includes the acquisition and honing of physical skills needed later in life, improving problem-solving abilities, cementing social relationships, and tuning the musculature and the nervous system (Martin & Caro, 1985). A notable feature of the mammalian nervous system is the superabundance of connections between neurons at the start of development. As the animal develops, many of these connections are lost and many cells die. Those neural connections that remain active are retained, and the unused ones are lost. This sculpting of the nervous system reflects the steadily improving efficiency of the body's classification, command, and control systems. These internal changes are reflected in behavior. When young animals playfully practice the stereotyped movements they will use in earnest later in life, they improve the coordination and effectiveness of these behavior patterns. The short dashes and jumps of young gazelle when they are playing bring benefits that may be almost immediate, as when they face the threat of predation from cheetah or other carnivores intent on a quick meal, and need considerable skill when escaping (Gomendio, 1988). The cheetah's own young also need to acquire running and jumping skills rapidly in order to evade capture by lions and hyenas (Caro, 1995). Even though the benefits may be immediate in such cases, they may also persist into adult life, not being lost in the behavioral metamorphosis that sometimes occurs during development.

Young animals may also familiarize themselves with the topography of their local terrain as a result of playing in it. Simply knowing the locations of important physical features will not guarantee rapid, safe passage around obstacles when escaping from predators or chasing prey. They need to practice. In keeping with this hypothesis, rats in a new area typically explore it first in a cautious manner. Gradually, the speed of movement increases until the animals are running rapidly around the area, along what become established pathways. The seemingly playful galloping ensures that, when fast movement becomes serious, the animal will be able to negotiate, efficiently and automatically, all the obstacles that clutter its familiar environment (Stamps, 1995). As it does so, it will be able to monitor the positions of predators, prey, or hostile members of its own species.

The argument continues that play allows young animals to simulate, in a relatively safe context, potentially dangerous situations that will arise in their adult lives (Smith, 1982). They learn from their mistakes, but safely. In this view, play exerts its most important developmental effects on risky adult behaviors such as fighting, mating in the face of serious competition, catching dangerous prey, and avoiding becoming someone else's prey. Indeed, the behavior patterns of fighting and prey catching are especially obvious in the play of cats and other predators, whereas safe activities, such as grooming, defecating, and urinating, have no playful counterparts.

If play is beneficial, then it follows that depriving the young animal of opportunities for play should have harmful effects on the outcome of its development. This is, indeed, the case. For instance, the lack of play experience shows clearly in the way the animal responds to social competition. In one experiment, young rats were reared in isolation, with or without an hour of daily play-fighting experience. About a month later each was put in a cage with another rat, where it was almost invariably attacked as an intruder. The defensive behavior of the play-deprived rats was abnormal. They spent significantly more time immobile than did animals that had played earlier in their lives. Other aspects of their defensive behavior were not affected, so the effects of play deprivation appeared to be specific (Einon & Potegal, 1991). It seems clear that such deprivation in early life adversely affected the animal's capacity to cope in a competitive world. The same argument may be mounted for play fighting in children. Through play, they learn how to cope with aggression and violence—their own and others'.

Distinguishing between the various hypotheses advanced to explain the current utility of play is difficult because the presumed benefits are usually thought to be delayed, appearing later in life, and developmental systems tend to be highly redundant, so that if an endpoint is not achieved by one route, it is achieved by another (Bateson & Martin, 1999). Playing when young is not the only way to acquire knowledge and skills; the animal can delay acquisition until it is an adult. But when such experience is gathered without play, the process may be more costly and difficult, even if it is not impossible. Play has design features that make it especially suitable for finding the best way forward. In acquiring skills, individuals (animals as well as humans) are in danger of finding suboptimal solutions to the many problems that confront them. In deliberately moving away from what might look like the final resting point, each individual may get somewhere that is better. Play may, therefore, fulfill an important probing role that enables the individual to escape from false endpoints.

Despite the uncertainties, play almost certainly has more than one biological function, even within a single species. Some aspects of play are probably concerned with honing the development of the nervous system and musculature. Other aspects are concerned with gaining an understanding of future social competitors and, if it comes to it, with the martial arts that will be needed to cope with them. Still other forms of play are involved with perfecting the predatory skills needed to catch prey without being injured, and some with developing efficient movement around a familiar environment to escape from predators or outwit competitors. Play is clearly not a *necessary* way for young animals to learn to recognize members of their social group, acquire knowledge of local culture, or become accustomed to their local environment; animals are patently able to acquire these forms of experience without playing. Nevertheless, these outcomes might still be beneficial consequences of play, if and when it does occur. They were not central to the evolution of play but, once it had evolved, any additional benefit was a bonus. The young animal is able to acquire, with no extra cost, information of crucial importance to it, such as recognizing close kin, in the course of playing for other reasons.

Although the functions of play are heterogeneous, the overarching theme of the functional argument is that the experience, skills, and knowledge needed for serious purposes later are acquired actively, through playful engagement with the environment. Human play has undoubtedly acquired yet more complex cognitive functions during the course of its evolution, rearranging the world in ways that ultimately enhance understanding. Creativity and innovation involve breaking away from established patterns. Creative people perceive the relations between thoughts, or things, or forms of expression that may seem utterly different, and are able to combine them into new forms—demonstrating the power to connect the seemingly unconnected. Play is an effective mechanism for facilitating innovation and encouraging creativity. Playfully rearranging disparate thoughts and ideas into novel combinations—most of which will turn out to be useless—is a powerful means of gaining new insights and opening up possibilities that had not previously been recognized. Play, in other words, extends to pure thought; it involves doing novel things without regard to whether they may be justified by a specified payoff.

THE ROLES OF PLAY IN EVOLUTION

Whatever the biological functions of play may be, the role of play in evolution raises quite different issues. The various ways in which the behavior of animals might have changed the course of evolution have become serious areas of inquiry. Four major proposals have been made for the ways in which an animal's behavior could affect subsequent evolution (Bateson, 2004). First, animals make active choices, and the results of their choices have consequences for subsequent evolution. Second, by their behavior, animals change the physical or the social conditions with which they and their descendants have to cope, thereby affecting the subsequent course of evolution. Third, by their behavior animals often expose themselves to new conditions that may reveal heritable variability and open up possibilities for evolutionary changes that would not otherwise have taken place. Fourth, animals are able to modify their behavior in response to changed conditions; this flexibility allows evolutionary change that otherwise would probably have been prevented by the death of the animals exposed to those conditions. Play might likely be involved in the fourth point, which is sometimes known as organic selection. I discuss this possibility in greater detail, but it is important to recognize at the outset that adaptations deriving from Darwinian selection may themselves involve considerable plasticity in the development of individuals (one of the most famous cases of this principle is behavioral imprinting; Bateson & Martin, 1999).

Although Douglas Spalding (1873), Bertrand Russell's tutor, probably had the idea first, modern thinking about the importance of behavioral plasticity in evolution stems from James Mark Baldwin (1896), Conwy Lloyd Morgan (1896), and Henry Fairfield Osborn (1896). Lloyd Morgan's account of the process was particularly clear and may be paraphrased as follows:

1. Suppose that organisms that are capable of change in their own lifetimes are exposed to new environmental conditions.
2. Those whose ability to change is equal to the occasion survive; they are modified. Those whose ability is not equal to the occasion are eliminated.
3. The modification takes place generation after generation in the

changed environmental conditions, but the modification is not inherited. The effects of modification are not transmitted through the genes.

4. Any variation in the ease of expression of the modified character that is due to genetic differences is liable to act in favor of those individuals that express the character most readily.
5. As a consequence, an inherited predisposition to express the modifications in question will tend to evolve. The longer the evolutionary process continues, the more marked will be such a predisposition.
6. Thus plastic modification within individuals might lead the process, and a change in genes that influences the character would follow; the one paves the way for the other.

It is obvious from this outline of the proposed process that Lloyd Morgan was not suggesting a genetic inheritance of acquired characters as a mechanism. The crucial postulate proposes a mechanism in the form of a *cost* of operating the original process of phenotypic adaptation—a cost that can be reduced subsequently by genotypic change, thereby enabling Darwinian evolution to occur. Alister Hardy (1965) did more than most in the intervening years to stress that the process could be of great significance. Hardy envisaged a cascade of changes flowing from the initial behavioral event. Even without structural change, control of behavioral development might alter over time. In order for this change to happen, adapting to the new conditions purely by trial-and-error learning or copying others must be more costly than doing it easily without active modification. One instance might involve differential responsiveness to particular types of food. Many cases in which a given species chooses an unusual food are probably not due to genetic changes but to the functioning of normal mechanisms in unusual circumstances. A group of animals might be forced into living in an unusual place after losing their way, but they cope by changing their preferences to suitable foods that are locally abundant. Later, those descendants that did not need to learn so much when foraging might be more likely to survive than those that could only show a fully functional phenotype by learning. A cost was incurred in the time taken to learn. As a consequence, what started as a purely phenotypic difference between animals of the same species living in different habitats became a genotypic difference.

The standard argument of those skeptical of organic selection is that if learning is highly beneficial in adapting to new conditions, why dispose of it? As Depew (2003) put it: "If learned behaviors are so effective in getting a useful trait passed from generation to generation at the cultural level, there will presumably be no selection pressure for the spread of genetic factors favoring the trait" (p. 15). I believe that this argument is based on an impoverished understanding of how behavior is changed and controlled. The answer to those who think that the proposed evolutionary change would lead to a generalized loss of ability to learn is to state quite simply that it would not. Learning in complex organisms consists of a series of subprocesses (Heyes & Huber, 2000). If an array of feature detectors is linked directly to an array of executive mechanisms as well as indirectly through an intermediate layer, and all connections are plastic (Bateson & Horn, 1994), then a particular feature detector can become nonplastically linked to an executive system in the course of evolution without any further loss of plasticity. Starting with an unsupervised neural net, it is easy to model organic selection, and the overall effects on the ability to learn new tasks are insignificant because of the massively parallel architecture of the brain (Bateson, 2004).

Since organic selection is often confused with Waddington's (1957) ideas about genetic assimilation, it is worth clarifying the differences. Organic selection involves necessary compensation for the effects of a new set of conditions and immediate response by the individual to the challenge. The accommodation is not inherited, and differential survival of different genotypes may arise from subsequent differences in the ease with which the new adaptation is expressed. Waddington's empirical findings also involved the expression of a novel character in a new environment, but the character was not an adaptation to the triggering condition, even though, in the regime of artificial selection used by Waddington, it conferred some advantage on its possessor. The novel phenotypes do not bear any functional relation to the conditions that disrupted normal development. Nor need there be such a relationship under natural conditions. All that is required initially is that the environmental conditions trigger the expression of a phenotype that can be repeated generation after generation, as long as the environmental conditions persist. The initial response of the animal in organic selection is fast, whereas the developmental effects of exposing animals to abnormal environments were not seen until they were adult. Since most individuals are adaptable, most will survive the initial stages of organic selection. In Waddington's experiments those expressing the novel character, a subset of the total population, were artificially selected for further breeding. Finally, in the case of organic selection described by Lloyd Morgan, fresh phenotypic variation presumably arises by a mutation that allows the adapted character to be expressed more easily and thence leads to differential survival. In Waddington's experiments mutation was neither postulated nor needed.

The existence of a phenotype, acquired by learning, sets an endpoint against which phenotypes that develop in other ways must be compared. In the natural world, if an unlearned phenotype is not as good as the learned one, in the sense that it is not acquired more quickly or at less cost, then nothing will happen. If it is better, then evolutionary change is possible. The question is whether the unlearned phenotype could evolve without the comparison. If learning involves several subprocesses as well as many opportunities for "chaining" (i.e., the discriminative stimulus for one action becomes the secondary reinforcer that can strengthen another), then the chances against an unlearned equivalent appearing in one step are very small. However, with the learned phenotype as the standard, every small step that cuts out some of the plasticity and simultaneously increases efficiency is an improvement.

As an example of how the setting of an endpoint might work, suppose that the ancestor of the Galapagos woodpecker finch (*Cactospiza pallida*) that pokes sharp sticks into holes containing insect larvae, did so by trial and error, and its modern form does so without much learning. In the first stage, a naive variant of the ancestral finch, when in foraging mode, was more inclined to pick up sharp sticks than other birds. This habit spread in the population by Darwinian evolution because those behaving in this fashion obtained food more quickly. At this stage the birds still learn the second part of the sequence. The second step is that a naive new variant, when in foraging mode, was more inclined to poke sharp sticks into holes. Again this second habit spread in the population by Darwinian evolution. The end result is a finch that uses a tool without having to learn how to do so. Simultaneous mutations that increase the probability of two distinct acts (e.g., picking up sticks and poking them into holes, in the case of the woodpecker finch) would be very unlikely. Learning makes it possible for them to occur at different times. Without learning, having one act but not the other has no value (Bateson, 2004).

If this line of thinking is correct, how could play have boosted the evolution of cognitive capacity? For a start, the tool-using cultures found in chimpanzees could have been present in the hominid lineage and then automated along the lines suggested for the tool-using cultures. In particular, those aspects of play that are creative or break out of local optima are especially promising candidates for driving evolution. The proposal is that when complex sequences of behavior develop, their components can be automated piecemeal by Darwinian evolution. Such improvements in what could be readily perceived as cognitive ability would not occur by genetic recombination or mutation, because the probability of the simultaneous occurrence of all the rare necessary events is vanishingly small. I have already rejected the commonly used objection that organic selection would lead to an overall loss of plasticity in development, on the grounds that learning processes are massively parallel in birds and mammals, especially in higher primates. The argument is, then, that aspects of play can, indeed, increase the total sum of spontaneously developing behavioral structure that serves to solve complex problems.

As with other aspects of evolution, no necessary link need exist between adaptation and the predictability of behavior. The evolutionary process can establish rules that affect how the individual changes its behavior in response to new conditions. To understand the opportunities that this regularity opens up, consider a rule-governed game such as chess. It is impossible to predict the course of a particular chess game from a knowledge of the game's rule. Chess players are constrained by the rules and the positions of the pieces in the game, but they are also instrumental in generating the positions to which they must subsequently respond. The range of possible games is enormous. The rules may be simple, but the outcomes can be extremely complex.

REFERENCES

Baldwin, J. M. (1896). A new factor in evolution. *American Naturalist, 30*, 441–451; 536–553.

Bateson, P. (2004). The active role of behaviour in evolution. *Biology and Philosophy, 19*, 283–298.

Bateson, P., & Horn, G. (1994). Imprinting and recognition memory: A neural net model. *Animal Behaviour, 48*, 695–715.

Bateson, P., & Martin, P. (1999). *Design for a life: How behaviour develops.* London: Cape.

Caro, T. M. (1995). Short-term costs and correlates of play in cheetahs. *Animal Behaviour, 49*, 333–345.

Deci, E. L., & Ryan, R. M. (1980). The empirical exploration of intrinsic motivational processes. *Advances in Experimental Social Psychology, 13*, 39–80.

Depew, D. J. (2003). Baldwin and his many effects. In B. H. Weber & D. J. Depew (Eds.), *Evolution and learning: The Baldwin Effect reconsidered* (pp. 3–31). Cambridge, MA: MIT Press.

Einon, D., & Potegal, M. (1991). Enhanced defense in adult rats deprived of playfighting experience as juveniles. *Aggressive Behaviour, 17*, 27–40.

Fagen, R. (1981). *Animal play behaviour.* New York: Oxford University Press.

Gomendio, M. (1988). The development of different types of play in gazelles: Implications for the nature and functions of play. *Animal Behaviour, 36*, 825–836.

Harcourt, R. (1991). Survivorship costs of play in the South American fur seal. *Animal Behaviour, 42*, 509–511.

Hardy, A. (1965). *The living stream.* London: Collins.

Heyes, C., & Huber, L. (2000). *The evolution of cognition.* Cambridge, MA: MIT Press.

Lee, P. C. (1984). Ecological constraints on the social development of vervet monkeys. *Behaviour, 91*, 245–262.

Lloyd Morgan, C. (1896). On modification and variation. *Science, 4*, 733–740.

Martin, P., & Caro, T. M. (1985). On the functions of play and its role in behavioral development. *Advances in the Study of Behavior, 15*, 59–103.

Osborn, H. F. (1896). Ontogenic and phylogenic variation. *Science, 4*, 786–789.

Smith, P. K. (1982). Does play matter?: Functional and evolutionary aspects of animal and human play. *Behavioural and Brain Sciences, 5*, 139–155.

Spalding, D. A. (1873). Instinct with original observations on young animals. *Macmillan's Magazine, 27*, 282–293.

Stamps, J. (1995). Motor learning and the value of familiar space. *American Naturalist, 146*, 41–58.

Waddington, C. H. (1957). *The strategy of the genes.* London: Allen & Unwin.

PART II

SOCIAL PLAY

CHAPTER 3

Social Play in the Great Apes

KERRIE P. LEWIS

Of ethological research conducted on mammalian play behavior, perhaps the largest concentration of reports is dedicated to the play of primates. Because primates are a typically social taxonomic order, social play behavior, or the play between two or more individuals, arguably receives the greatest attention. Furthermore, because great apes are known to be intelligent and are widely seen as especially charismatic among the primate order, their social play behavior has been of special interest to ethologists.

Both captive and wild chimpanzees (*Pan troglodytes*), bonobos (*Pan paniscus*), gorillas (*Gorilla gorilla*), and orangutans (*Pongo pygmaeus*) are noted as being especially playful animals. Social play, in particular, constitutes some of the most colorful and diverse behavior seen in the primate world, taking many forms. Social play behavior has recently been proffered as a behavioral category distinct even from other forms of play (Panksepp, 1998). Indeed, observation of social play may offer significant insights into the lives of the animals that perform it, in terms of innovation, social affiliations, dominance relationships, cultural transfer, and cognitive capacities. Using examples from the literature, this chapter offers a comprehensive overview of social play in the great apes within the context of play within the primate order.

THE DIFFICULTY WITH PLAY

Attempts to define play behavior in any capacity have been historically tricky (e.g., Bekoff & Byers, 1981). As humans, we seem to be able to instinctively recognize play in humans and animals, but it is often difficult to describe what it is that enables such recognition (e.g., Bekoff & Byers, 1985, 1998). The study of play has been traditionally problematic because play "borrows" its behavioral elements from other behaviors, such as sex and aggression. That is, the behavioral repertoire comprising play also encompasses behaviors that we might usually be more likely to assign to other behavioral contexts, such as agonistic encounters (e.g., Chalmers, 1984). With this caveat in mind, it has often been said that play is defined by what it is not; thus, all too often it becomes a default explanation for a behavior that appears functionless (Martin & Caro, 1985). Consequently, several decades of play research have seen considerable debate in the literature as to what elements constitute play (Allen & Bekoff, 1997; Fagen, 1981). To avoid becoming entangled in the semantics of such a debate, I refer the reader to Bekoff and Byers's (1981, p. 301) now widely cited definition that provides an excellent basis by which to understand play simply:

> Play is all locomotor activity performed postnatally which appears to an observer to have no obvious immediate benefits for the player, in which motor patterns resembling those used in serious functional contexts may be used in modified terms. The motor acts constituting play have some or all of the following structural features: exaggeration of movements, repetition of motor acts, and fragmentation or disordering of sequences of motor acts.

Animal recognition of playful interactions is usually somewhat automatic or instinctive in response to the apparent relaxation of sexual or aggressive "intent" in play. That is, although two individuals may be engaged in a bout of wrestling, there is a marked lack of any true aggression. It might be fair to suggest that a continuum exists between play behavior and more "serious" behaviors, especially given that the boundaries of play become increasingly blurred as a juvenile approaches adulthood, when it becomes difficult to differentiate play fighting from actual aggression if play fights end antagonistically (Paquette, 1994). For this reason, play is widely regarded as the behavioral domain of infants and juveniles, and its development is of particular interest to play ethologists.

CATEGORIES OF PLAY BEHAVIOR

The classification of play typically falls into one of three categories: solitary locomotor–rotational, object manipulation, and social play (Bekoff & Byers, 1981, 1998). Social play has several subcategories, the most widely observed being that of play fighting or rough-and-tumble play, which involves chasing and wrestling behaviors. Certain aspects of juvenile allomothering behavior, such as carrying and cradling other infants, are sometimes referred to as "play-mothering" (Lancaster, 1971), although some workers disagree with this classification as play behavior (Smith, 1982). Additionally, the mounting behavior of sexual play forms part of the play repertoire of the great apes, and is typically thought to promote locomotor and social skills and to provide sexual experience (Vankova & Bartos, 2002). Juveniles typically cannot reproduce, and so sexual play behavior is sometimes referred to as "pseudosexual play." Reports in the literature suggest that the exhibition of pseudosexual behavior, in general, is higher in captive apes than in wild apes, and that these behaviors may be directed toward an adult rather than a juvenile conspecific (Spijkerman, van Hooff, Dienske, & Jens, 1997). The great apes are unique in that they have been reported to engage in almost all of the nonhuman primate play behavioral repertoire, including rough-and-tumble play, play-mothering, pseudosexual play, and adult play interactions. The behavioral repertoires of ape social play are detailed in Table 3.1.

TABLE 3.1. Ethogram of Social Play Behaviors in the Great Apes

Social play category	Typical characteristics
Play fighting	*Aggressive movement patterns without threat gestures*: wrestle; slap; jump at; jump on; play bite; cuff; pull; rough-and-tumble; roll; run toward; kick; sniff; chase; attack–withdraw; spar; drag; pinch; hit
Other	*Less aggressive playful social interactions*: finger wrestle; peek-a-boo; gentle roll; grapple; tickle; play walk; laugh; hair tug/pull
Adult	*Usually rare; usually with infant (offspring)*: tickling; gentle wrestling; peek-a-boo; nibble; grapple; embrace; sham bite; dandle; head bob; head shake; grasp; lift
(Pseudo-) sexual	*Adult; subadult; older juveniles*: mounting; genital inspection; licking; rubbing; thrusting
Play-mother	*Juveniles, often female*: groom; carry; inspect

PLAY SIGNALS

How does a player or an observer distinguish between play and other categories of behavior? Agonistic signals can occur during play fighting, rendering the distinction between play and other behaviors that much harder to accurately recognize. It seems that it is not so much the actual behavioral patterns that are used but the way in which they are performed that is important, and this is where distinct play signals may come into operation. Because it would be especially risky for an animal to misinterpret the intention of any behavior pattern, play signals have evolved to distinguish serious from nonserious behavior patterns, allowing an appropriate and constant response to another individual (Bekoff & Allen, 1998). Play cues are often seen in their more functional contexts in other forms of behavior.

The play face, especially common to primates, typifies such a play signal: the relaxed open-mouth display that appears in playing primates, carnivores, and rodents (Fagen, 1981; Figure 3.1). In chimpanzees, a play face is often accompanied by a vocalization that approximates, but is distinct from, human laughter (Marler & Tenaza, 1977), and baboons often emit a play chuckle (D. Custance, personal communication, July 1999).

Elsewhere in the animal kingdom, rats are reported to produce chirps as possible laughter in response to being tickled (Panksepp & Burgdorf, 2000), and many carnivores also emit play vocalizations to signal a motivation to play (Bekoff, 1974). Although these vocalizations persist through certain mammalian orders, much social play is typically silent, traditionally thought to avoid attracting the attention of predators; more recently, how-

FIGURE 3.1. Chimpanzees in play. Note open-mouthed play face. Photograph by Kerrie Lewis (Singapore Zoo).

ever, a phylogenetic connection between play vocalizations and the occurrence of allo-mothering behavior has been shown (Masataka & Kohda, 1988). Play calls are likely to function to signify intention to play, but they may further have mechanisms as a source of information for infant retrieval. Play signals may have their behavioral basis in other functional categories of behavior, such as fighting or mating, yet they are usually exaggerated actions and thus unique to play behavior (Fagen, 1981). Play signals are not necessary for play to occur, but their inclusion might aid in initiating and maintaining a play bout, which play ethologists often contend is instrumental to the "success" of play (Watson, 1998).

SOCIAL PLAY AND LAUGHTER

One such play signal is that of laughter. In humans, laughter is often contagious, and occurs frequently in social contexts. This remains true for the great apes, who exhibit play-specific vocalizations that may be akin to what we as humans term "laughter."

Chimpanzee laughter has been discussed since the days of Darwin (1872); Dian Fossey (1979) reported gorillas "chuckling." A study by Förderreuther and Zimmerman (2003) on captive bonobos demonstrates that infant bonobos and infant humans share a very similar pattern of facial expressions and vocalizations during playful interactions. The authors argue that because these patterns are so common across the great apes, it is likely that laughter was present in the common ancestor of apes and humans, emerging as a signal to demonstrate to others that an encounter was playful rather than aggressive.

Indeed, both wild and captive chimpanzees are reported to exhibit panting sounds during play that resemble laughter, but whereas humans tend to laugh during noncontact social situations, chimpanzees laugh almost exclusively in the context of playful contact. Additionally, human laughter is produced by "chopping" a single exhalation, whereas chimpanzees pant in and out. This difference may further suggest a constraint on great ape spoken language ability. Language constraints aside, laughter is almost certainly a social signal that evolved within social contexts, and social play is perhaps one of the most important behavioral categories in this vein (Kipper & Todt, 2001).

Smiling is widely considered to have originated as an ancient mammalian threat display, indicating the potential for harm or self-defense; it may be the case that a simple flash of teeth evolved as a social signal to demonstrate this intent and required no further action (Andrew, 1963). Laughter, however, is shown to have a different origin and may be more keenly founded in the ancestral emergence of play (Panksepp, 1998). Laughter may act as a public and social demonstration of victory (e.g., winning a play fight) or to promote social cohesion (e.g., human laughter is "infectious"). Neurologically, the gap between laughing and crying is small. It is hypothesized that the mechanisms that permit the experience of separation–distress, along with those that promote the emotional experience of social bonding, may have been prerequisites for the origin of laughter (Panksepp, 1998). If this is the case, it seems intuitive that the mechanisms allowing the experience of social bonding and separation–distress, as well as the propensity to cry and to laugh, may well have been prerequisites in the neural circuitry for the evolution of play.

Although scientists have to be especially vigilant in avoiding anthropomorphism, and especially primatocentrism, it is hard to escape the apparent exhibition of similar emotive states in great apes, especially given their proximity to humans as our closest living relatives. Social cohesion for group-living species is vital, and social interactions, such as social play, that help to facilitate social group living, are important in maintaining this cohesion. Social signals, such as smiling and laughing, are likely to have evolved in socially cohesive groups, indicating and allowing the safe continuation of close sociophysical contact.

AGE AND SEX DIFFERENCES

Data on age-specific rates of play are rare. Nevertheless, there are two main hypotheses about play behavior: that juveniles play more than adults, and that males play more often and more roughly than females (*chimpanzees*: King, Stevens, & Mellen, 1980; Markus & Croft, 1995; *gorillas*: Brown, 1988; *orangutans*: Nash, 1993). That play is less common in older adults may be due to the fact that motor skills are not required to the same extent in older compared to younger animals (Watson, 1993). If one type of play behavior were riskier for females than for males, for example, then it would be expected that female play behavior would decrease.

As an animal matures and develops, even within relative immaturity, it often becomes harder to distinguish true aggression from boisterous rough-and-tumble play (Martin & Caro, 1985). It certainly seems to be the case that an individual primate's capacity for play varies according to its age and sex (Chalmers, 1984). It seems that dominant animals are capable of moderating their play behavior according to relevant asymmetry in strength; if one juvenile plays too roughly with another, it risks alerting the attention of the mother, who may break up the play interaction. As primates mature, they become more aware of intratroop status and more competitive in play (Fagen, 1981). Perhaps this is a reason why older male juveniles play more roughly than any other age group (Owens, 1975; Rowell & Chism, 1986; Schåfer & Smith, 1996).

It has been widely documented that juvenile males, in particular, become increasingly aggressive with age (Kraemer, Horvat, Doering, & McGinnis, 1982; Pusey, 1990). This finding may help us understand why juvenile males appear to play more frequently, and also more roughly, than their female counterparts. In sexually dimorphic mammals, females mature earlier than males, which may have some effect on sex differences in play behavior. In orangutans, older juvenile females tend to play less frequently and exhibit less aggressive play than their male counterparts, concentrating more on interactions with infants and alliances with other females. Males of the same age, however, do not form alliances in the same way, but concentrate on rough play with other males—a behavior that may, in part at least, serve as a practice of fighting skills (Nash, 1993).

Chimpanzees have not been reported to exhibit significant sexual dimorphism in play (Fagen, 1981), but sex differences have been observed in many wild, captive, and laboratory studies on primate play behavior (e.g., *Gorilla gorilla*: Brown, 1988; Meder, 1990). Self-handicapping by older or stronger play partners may often occur to maintain a play bout with a mismatched partner (Pereira & Preisser, 1998). However, it has been shown empirically that primates predominantly prefer to choose same-age, same-sex, and same-rank play partners (e.g., *Cercopithecus aethiops sabaeus*: Govindarajulu, Hunte, Vermeer, & Horrocks, 1993).

DEVELOPMENTAL TRAJECTORY

The developmental trajectory of play during the juvenile period typically shows two peaks in play frequency: one soon after play begins in infancy, and one later in the juvenile period. The first peak is usually followed by a rapid decline in play behavior, which eventually increases to bring about the second peak in play activity just prior to weaning age. After this period, play declines steadily with the onset of adulthood. Indeed, the patterning of play during the juvenile period has been used to hypothesize that mammalian play has evolved to enhance the developing neuromuscular system during sensitive periods (see below, and also Byers & Walker, 1995; Fairbanks, 2000). Figures 3.2 and 3.3, taken from Lewis (2003), detail the typical developmental trajectory of play in gorillas and chimpanzees over the first 40

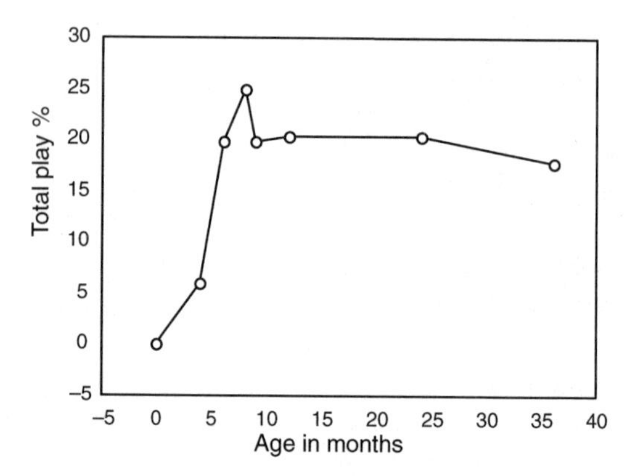

FIGURE 3.2. Developmental trajectory of play in gorillas. From Lewis (2003). Reprinted by permission of the author.

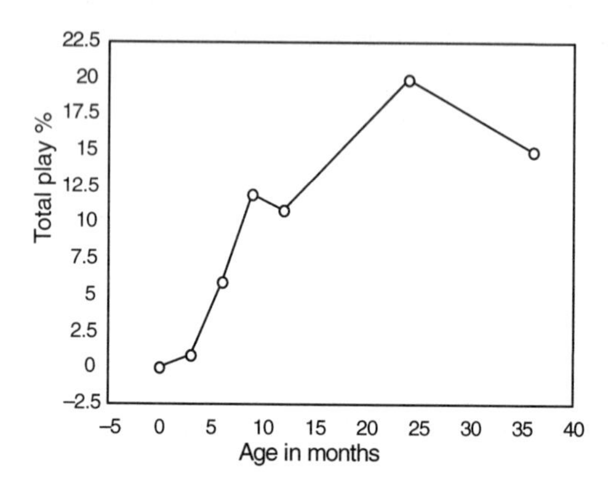

FIGURE 3.3. Developmental trajectory of play in chimpanzees. From Lewis (2003). Reprinted by permission of the author.

months of life. Age at weaning is 52 months in gorillas and 48 months in chimpanzees.

ADULT PLAY

Although play behavior is most usually associated with infants and juveniles, adults of some species have been reported to engage in it, both with other adults and, of course, with their infant offspring, and the great apes are no exception (*chimpanzees*: Bloomsmith, 1989; *bonobos*: Enomoto, 1990; *gorillas*: Fischer & Nadler, 1978; *orangutans*: Zucker, Mitchell, & Maple, 1978). Great ape adult play appears in various forms: solitary adult play, adult–adult play and adult–juvenile play. Adults may play with their offspring or with other juveniles within the group, and mother–infant play is common in most primate species. Although play is relatively common between a mother and her offspring, it is often the case that other adult members of the group, including males, might engage in playful activity with youngsters (*orangutans*: Zucker et al., 1978; *bonobos*: Enomoto, 1990; *chimpanzees*: Pruetz & Bloomsmith, 1995).

Adult–juvenile play may be adaptive, and it has been noted that in species where both parents assist in the rearing of offspring, playing offspring might benefit the parent in terms of gene persistence. Yet we also observe play between adult males and infants within promiscuous species, such as chimpanzees. If it is assumed that the function of adult–infant play is to assist the physical and behavioral development of offspring (Fagen, 1981), then we might only expect to see such play interactions if an adult male can be somewhat sure that the infant with whom he is playing is, in fact, his own offspring (unless he wants to harm the infant). Moreover, adult males who can be comparatively sure of their paternity are more likely than other males to protect and care for infants in any sense (Hrdy, 1976). However, in captivity at least, it has been shown that there are no significant differences between the play of adult male chimpanzees and their likely offspring and the play that occurs between adult males and other-sired offspring (see Pruetz & Bloomsmith, 1995); captive adult orangutan males may play with their offspring, although this is rarely observed in the wild (Zucker et al., 1978).

Parental play may have certain benefits of assisting the development of offspring and maintaining aspects of their own social relationships (Fagen, 1981). In fact, if social play helps facilitate the practice of social skills used later in life, then an adult's play with its offspring may arguably represent a form of parental investment (Fagen, 1981; Zahavi, 1977). Moreover, it has even been suggested that the exhibition of play behavior offers a means by which parents can assess the behavioral and physical competence of their offspring (Chiszar, 1985), and indeed rough-and-tumble play may function to perfect social competence and affect the transition from juvenile to adolescent (*humans*: Pellegrini, 1995). The degree of relatedness between players certainly seems to be important within adult play behavior. It is more likely that an adult's play partners, regardless of age, will be close kin (or potential mates); relatedness is also important for juvenile players, in that play partners are often selected on the basis of the social relationships of the player's mother (*chimpanzees*: Tomasello, Gust, & Evans, 1990).

SOCIAL PLAY

Play is a behavior that begins very soon in postnatal life. Among great apes and humans, social play is arguably the first to emerge through playful mother–infant interactions, and it quickly becomes the dominant form of play behavior as young apes continue to playfully interact with their mothers, and also their kin and peers, growing in strength and boldness (Fairbanks, 2000). Social play is, without doubt, the most common category of play behavior seen among primates, and the most widely reported, both due to the prevalence of social play over solitary and object play, and also because play is most easily distinguished from other forms of behavior when it is social in nature.

Although the performance of play diminishes with the increased stress of suboptimal conditions such as predator pressure, inclement weather, and reduced food resources, young animals strive to perform play against all possible odds (Fagen, 1981). Time spent playing in mammals has been reported as between 1 and 10% of the daily time budget for most species (Fagen, 1981), and among the great apes, social play is by far the most commonly performed category of play behavior. In an extensive comparative study of play behavior in primates, Lewis (2003) found that the average time budget allocation for social play, based on the published literature, was around 14% for each of the great ape species. This is compared to an average of nearer 8% among other haplorhine primates. The great apes typically show significantly greater amounts of social play than any other primate group.

Social play is typically considered the most "cognitive" form of play behavior, because in order to be successful, social players need to pay attention to the behavioral cues of their play partners and react quickly, accurately, and appropriately (Špinka, Newberry, & Bekoff, 2001). Misjudging the behavioral rules of social play poses considerable risk in terms of potential injury and social distress—not the least of which is that an inappropriate response renders an immediate end to the play bout (Martin & Caro, 1985). Reacting to social cues in order to persist in the social play bout must require increased cognitive processing than the performance of nonsocial play. Given that the great apes are known to have especially large brain sizes relative to other primates, it is perhaps unsurprising that they are the most socially playful of all primates. This pattern remains true of other large-brained orders, such as cetaceans, as well.

SOCIAL PLAY FUNCTIONS

Many theories have been proffered as to what an animal gains by playing, yet it is not always apparent what the functions of play might be (Bekoff & Allen, 1998). Table 3.2 details some theories of the different functions of social play. Early research viewed social play as a means by which an animal is able to practice skills in readying for adulthood. Although elements of this hypothesis remain undoubtedly true, more recent research places the primary function of social play not as mere practice but as a mechanism by which the animal engages in complex behaviors, thus giving the body and brain experiences in working together and in responding to fluctuating and unpredictable conditions (Burghardt, 2001; Špinka et al., 2001).

As noted, just as juveniles are more playful than adults, males of any given species are typically more playful than their female counterparts (Jolly, 1985). Indeed, testing the boundaries of individual dominance is an important factor in the lives of male chimpanzees as they strive to attain social control (deWaal & van Hoof, 1981). One method for safely experimenting with such skills is through social play, which provides practice and tests the limits of the young male's own strength, agility, social placement, and playful deception directly against conspecifics through play fighting, wrestling, chasing, and other playful behaviors. Spijkerman and colleagues (1997) point out that teasing is one such playful behavior that permits the exploration of social boundaries through testing the uncertainty of social responses in conspecifics.

Considering the time frame during which social play is most prevalent—that is, almost exclusively during the sensitive juvenile period of development—social play may represent a developmental marker for per-

TABLE 3.2. Possible Functions for Social Play Behavior

Social play element	Proposed function	Sources
Social play in general	For cognitive enhancements of social interactions and in "reading" unpredictable social situations	Špinka et al. (2001)
	To relieve group tensions	Enomoto (1990)
Social play signals	To demonstrate or practice intent	Bekoff (1975, 1977, 2001a)
Play fighting	To practice dominance and aggression/test strength/mate defense	Bramblett (1976); Fagen (1981); Pellis & Pellis (1996)
	To develop communicative abilities, especially during agonistic situations	Dolhinow (1971)
	To demonstrate behavioral and physical competency	Chiszar (1985)
	To learn cooperation/trust/"fair play"	Bekoff (2001a, 2001b)
Self-handicapping	To create practice for the "unexpected"	Špinka et al. (2001)
Play-mothering	To develop/practice maternal behavior	Lancaster (1971); Meaney et al. (1985)

manent changes to take place in the central nervous system, effectively "wiring up" the developing brain. Thus, play may assist, or at least contribute to, the development of permanent effects on physical, neural, and mental processes (Byers & Walker, 1995; Fairbanks, 2000; Lewis, 2003; Lewis & Barton, 2004). It is worth noting, however, that although social play may aid the learning and experience of some social interactions, not all learning occurs through play, and animals deprived of play as juveniles may still develop normally (Martin & Caro, 1985).

Play is sensitive to environmental constraints. In both wild and captive settings, external factors such as temperature, weather, food availability, and predator pressure can have negative effects on the amount of play exhibited (e.g., Burghardt, 1984). Perhaps unsurprisingly then, it seems that the social environment provides a hugely important underpinning for the exhibition of social play. Socially isolated animals typically fail to show significant signs of playfulness. Classic studies demonstrate that relative to wild-raised, and even other captive-raised, chimpanzees, socially isolated infants do not exhibit normal social behaviors during juvenility and adolescence, including social play, even when reintroduced to a social group. Two decades ago, Smith (1982) pointed out that play fighting is not a prerequisite for real fighting, and that adult "serious" behaviors can and do occur without the early experience of play. Indeed, this finding is not surprising, given that not all adult learning occurs through play (Martin & Caro, 1985).

However, animals that are socially isolated during infancy, even if resocialized, engage in social play far less than normally housed infants and instead exhibit augmented aggressive and fearful behaviors (Ruppenthal, Arling, Harlow, Sackett, & Suomi, 1976; Suomi & Harlow, 1972). From similar studies, it seems that the quality of the social group during juvenility has much bearing on the social behaviors exhibited: for example, chimpanzees raised in peer groups (age-mates only) demonstrate lower levels of normally sexually differentiated behavior in their social play, lower levels of dominance and even general activity, but higher levels of stereotypic behavior, in comparison with chimpanzees raised in larger and more diverse social groups (Bloomsmith, Pazol, & Alford, 1994; Spijkerman et al., 1997). Thus, engaging in social play at an early age may benefit the individual in the acquisition of social competence and confidence. Given the typical prevalence of social play to the behavioral repertoires of the great apes, it is likely that the skills acquired through social play behavior are instrumental in developing and maintaining social relationships and bonds that may last a lifetime. Indeed, the skills acquired in facilitating such social relationships through the performance of social play behavior certainly benefit the individual later in adulthood, when successful social relationships become all the more vital.

SOCIAL PLAY IN THE GREAT APES

Great ape play shares many features in common with the play of other primates. The following sections discuss which features unite and divide social play in each of the four great apes.

Gorillas

Gorillas are the most sexually dimorphic of the great apes, with adult males weighing up to 180 kilograms and females weighing around 75 kilograms (Rowe, 1996). As such, marked social play preferences are also apparent, with males demonstrating a strong preference for animated play fighting and females partaking in quieter social play patterns that feature more play-mothering and play-nesting behaviors, as well as increased allo-grooming, in comparison with males (Meder, 1990). Gorilla social play largely involves running, wrestling, and chasing (Freeman & Alcock, 1973; Meder, 1990; Schaller, 1972). Males are usually the initiators of play, grabbing, charging at, or slapping their chosen partner in an attempt to begin a play bout. In heterosexual play partnering, males are usually more likely to both initiate play and chase their female play partners, than vice versa (Meder, 1990), and perhaps this one-sidedness is why a preference for same-sex partners for both males and females becomes apparent. Schaller (1972) noted that over 80% of infant social play bouts are dyadic (two players), and some 20% are polyadic (three or more players). Thus gorilla social play can be highly complex, given the sociocognitive abilities required to maintain play among larger numbers of players.

Like all species of primate, gorillas exhibit play faces during playful interactions; however, Tanner and Byrne (1993) reported that a female gorilla sometimes hid her play face with her hands. When the gorilla did this, she did not engage in play behavior, despite the appearance of other play signals. When her play face was exposed, normal play followed. Explanations of this behavior are unclear, although it suggests that the gorilla is acutely aware of her spontaneous facial expressions and has sufficient understanding of this fact to suppress the expression and change her behavior accordingly.

Chimpanzees

Chimpanzees play first with their mothers; early mother–infant social interactions predominantly involve grooming and play. During early infancy, the mother and infant begin interacting with other conspecifics, largely females, and by 2 years old, when the infant is somewhat weaned, they play less with their mothers, having formed functioning socially interactive play relationships with other infants and juveniles (Horvat & Kraemer, 1981). As juveniles, chimpanzees appear to play indiscriminately with both kin and non-kin peers, but as male juveniles approach adolescence and early adulthood, play with non-kin generally ceases, although play with kin may persist (van Lawick-Goodall, 1968).

Younger chimpanzees have the most to gain through playing with as many peers as possible. However, as the time budget for social play naturally decreases with age, it is possible that this time is more beneficially spent aiding the physical and social development of related individuals, than among those who do not share genetic relatedness. As young chimpanzees approach adulthood, their time spent in play increasingly becomes replaced with another important social behavior: allo-grooming (Merrick, 1977). These two behaviors are structurally, and in many ways, functionally, different. However, the positive effects of social cohesion in chimpanzees, which accrue through both social play and social grooming, are noted in the literature (Merrick, 1977).

Chimpanzee social play is noted as being especially colorful, often involving objects and elaborate chasing and wrestling games. In the wild, chimpanzees have been noted for their use of tools in termite fishing and nut cracking; indeed, among Gombe and Mahale chimpanzee troops, tools are often used to initiate play bouts (McGrew, 1996). Captive chimpanzees are typically more playful than their wild counterparts, especially during adulthood. Van Lawick-Goodall (1968) stated that adult females might engage in play with other adults maybe two or three times a year; this number increased during times when the Gombe troop were artificially provisioned, suggesting that a decrease in stress associated with foraging permitted an increased level of play behavior. Captive chimpanzees of all ages play in excess of this level.

Bonobos

In comparison with the more familiar common chimpanzee, the behavior of the pygmy chimpanzee, or bonobo, is less well understood. However, a detailed study of social play in bonobos by Enomoto (1990) showed that these apes also exhibit behaviorally complex social play behaviors from all categories of social play, including adult male–infant play. DeWaal (1995) describes the innovative games played by juvenile bonobos, including those with apparent "rules" such as "blindman's bluff," in which players use hands, arms, or even banana leaves to cover their eyes while flailing around their habitat, attempting to find their play partners.

Bonobo social behavior is especially noted for its high sexual content, with frequent copulations occurring between all group members, even when age and sexual cycle cannot permit conception (Hashimoto & Furuichi, 1996). Frequent sexual contact takes place during instances of interest–arousal, such as feeding; additionally, if a novel object is thrown into a captive bonobo enclosure, sexual mounting might occur before the bonobos continue to play together with the object (deWaal, 1995). These sociosexual behaviors are believed to enhance the social cohesion of the group by strengthening social relationships between group members, aiding reconciliation, reducing tension, and diffusing aggression (deWaal, 1995). It perhaps comes as little surprise that much social play in bonobos is sexual in appearance. Genital contacts are frequently seen in play-like contexts, especially between immature individuals; and, like all play behavior, sexual play among immature bonobos is more common in males than in females (Hashimoto & Furuichi, 1996). Although genital contact is common in the play of bonobos, its appearance is virtually absent in infants, then emerges and increases with age, ultimately replacing social play behavior as a bonobo reaches adulthood (Hashimoto, 1997).

Orangutans

In comparison with African apes, much less is known about the behavior of orangutans, because they do not form the large social groups that we see in chimpanzees, bonobos, and gorillas. Instead, orangutans are more arboreal and solitary, with primary social groupings being between a mother and her offspring, and with males joining these small groups during female sexual receptivity. As a result, orangutans neither need nor exhibit the raucous social play episodes witnessed in the behavioral repertoires of African apes. In contrast, they are far quieter, both vocally and in physical exuberance (Harrison, 1961, 1962). However, whereas adult males typically do not form part of everyday social groupings, captive studies have shown that, given certain conditions, adult males do engage in social play with their offspring (Zucker et al., 1978).

In a unique study that compared jointly housed juvenile gorillas and orangutans, Freeman and Alcock (1973) described qualitative and quantitative differences in the play of these species, with orangutans engaging in leaping, swinging, and dangling displays, and gorillas in climbing and running-based play. Interestingly, the orangutans' social play was marked by their tendency to engage in hiding games. This difference may reflect the differing social ecologies of orangutans compared with gorillas.

All great ape species build nests, usually for the prime purpose of resting and sleeping. One new finding about orangutan social behavior is that the orangutans build nests specifically for the purpose of social play (Van Schaik et al., 2003). This practice is somewhat different from other apes. For example, although bonobos may also engage in social play behaviors within their nests, the proportion of time in nests devoted to social play is about 3%; the remainder of time in the nest is devoted to resting, grooming, and feeding (Fruth & Hohmann, 1996).

SOCIAL PLAY AND COGNITIVE EVOLUTION

Although understanding the form and function of social play is important, it is vital to study these areas in relation to behavioral evolution if we are to achieve a deeper knowledge of social relationships. Social play is a behavior seen across many species, and as such, has a strong evolutionary history. Social play certainly occurs in all ape species and in most other primate species, but it is unlikely to represent the earliest form of play behavior, since sociality is not a primitive condition in mammalian evolution. Byers (1984) states that play patterns that resemble the serious behavioral components of the flight response are the most likely to have evolved first; thus, solitary locomotor–rotational play behavior is the most likely representation of the earliest form of play, followed by play that mimics adult agonistic competition (social play).

Social play is frequently deemed a separate category of play behavior because of the cognitive and social skills required to perform it successfully, as discussed above. It is widely proffered that social play evolved in association with social group living. Indeed, it is likely that social play arose with neocortical expansion (Lewis, 2000), given that neocortex size can be used to predict measures of sociality (Dunbar, 1992), and that neocortical wiring is critically determined by experience early in life (Quartz & Sejnowski, 1997). Presumably, with increasing neocortical expansion emerged the specialization for sociality, which in turn facilitated the divergence from the primitive play condition toward the diversity and complexity of social forms of play. Future studies should seek to demonstrate to what extent this might be the case.

The neural basis for the development of skills through play still remains comparatively unexplored. Indeed, fairly little is actually known about the developing brain during the early and late phases of the juvenile period, a time during which social and cognitive learning escalates and play is at its most prominent (Casey, Giedd, & Thomas, 2000). Iwaniuk, Nelson, and Pellis (2001) found that at an interspecies level in primates, there was no correlation between play and total brain size. Lewis (2000) found that in primates, social, but not other forms of play behavior, correlate significantly and positively with neocortex size across taxa. It might be expected that neocortex development is important to the onset of play behavior, because this development begins, and is strongest in, the early postnatal phase (Dąmbska & Kuchna, 1996). It is in this early postnatal period that play is especially prevalent. Dunbar (1992, 1995), and Dunbar and Bever (1998) have demonstrated a relationship between neocortex size and group size in primates and carnivores, arguing that the larger the social group, the greater the cognitive load in maintaining social relationships.

Although comparative studies of cognition have focused exclusively on the neocortex, it has recently been established that the neocortex has tended to evolve together with the cerebellum (Barton & Harvey, 2000; Whiting, 2002—but see also Barton, 2002; Clark, Mitra, & Wang, 2001), suggesting that the cerebellum ought to be given more attention. It is likely that different parts of the brain are involved in different aspects of play behavior, but the extent of this participation is currently unclear. Increasing evidence shows that motor development and cognitive development are much more tightly interrelated than was previously thought (Diamond, 2000; Joseph, 2000; Willingham, 1999); other recent research has shown that social play coevolved with certain aspects of brain function. For example, Lewis and Barton (2004) demonstrated a coevolutionary relationship between the cerebellum and social play, suggesting a link between both cognitive and motor aspects of playful interactions (see also Lewis & Barton, 2004; Siviy & Panksepp, 1987). Indeed, social play is further argued to enter into the transfer of cultural skills, such as tool use, and it is often used to determine dominance rank and social relationships, including possible paternity (Colvin, 1983). In addition, the great apes' use of various gestures and facial expressions in their social play may offer some insight into the evolution of smiling, laughing, and arguably, even humor (Kipper & Todt, 2001; Panksepp & Burgdorf, 2000; Ramachandran, 1998).

SOCIAL SKILLS AND SOCIAL INTELLIGENCE

The social play of animals is likely to have an influence upon other social skills pertaining to individual functioning within the social group, allowing members to learn about and maintain social relationships (Lee, 1983; Mendoza-Granados & Sommer, 1995). Joffe (1997) suggests that primates are selected for an extended juvenile period, and that this characteristic is presumably related to their need to learn. One means of learning adult social skills is through play behavior, and the extended juvenile period in primate development suggests that increased play behavior occurs during this period to maximize learning potential.

Play behavior in juveniles may be a factor in the formation of dominance hierarchies and social ranking in some species, such as chimpanzees (Bramblett, 1976). Indeed, there also might be a clear preference to play with individuals of a similar dominance rank, or with infants of mothers who are high ranking (Berman, 1983; Colvin, 1983; Koyama, 1985; Lee, 1983). As an infant slowly matures, it spends increasingly more time away from its mother, forming peer bonds and engaging in playful interactions with individuals of similar age and rank (Berman, 1983; Govindarajulu et al., 1993). Initially, there is no injury as a direct result of play, but as the animals grow in maturity and strength, incidences of injury increase. It is argued that the individuals who are playful but injured less frequently may acquire a higher social rank than the possibly weaker individuals who sustain more injuries or who retire from playful interactions with their peers (Bramblett, 1976, p. 36; deWaal, 1996). In this way, play may have a direct function in establishing rank relationships.

However, it has been argued that allo-grooming is a better indicator of social bonds than social play behavior, especially because it is less energy consumptive (Colvin, 1983; Poole, 1985). In addition, dominance hierarchies may be decided through direct threat and submissive behaviors; thus play fighting, for instance, might not be used in this pursuit due to the relaxed nature of play and the self-handicapping tendency of these primates (Humphreys & Smith, 1987; Pellegrini & Smith, 1998; Pereira & Preisser, 1998). Furthermore, there is evidence to suggest that social play does not underpin group socialization: Some solitary animals demonstrate rather complex social play episodes with siblings or nest mates (e.g., *orangutans*: Rijksen, 1978; Zucker et al., 1978). Similarly, some social species (e.g., mice: *Mus musculus*) are not observed to play at all (Poole, 1985; Poole & Fish, 1975). So play may aid only some aspects of socialization, the formation of bonds, and information gathering regarding strength and dominance (Bekoff, 1978; Bramblett, 1976), although this might not be the case universally in mammals. The question of play as a vehicle of socialization further indicates the functional enigma of play behavior.

Generally, juveniles are unable to reproduce and thus usually unable to contribute genes to the next generation. It may be the case that developing skills and relationships through play to aid survival into adulthood will increase their lifetime fitness (Janson & Van Schaik, 1993, p. 57). Adult skills, or the pathways leading to them, need to be acquired at some point prior to attaining adulthood, and so it seems probable that play may aid, at least in part, in the development and learning of such skills (Fagen, 1993). Play and learning are important components of developmental patterns and may be especially useful in the practice of mating and mothering behavior patterns. Play-mothering in juvenile females may allow for the practice of behavior patterns and skills utilized in motherhood. Thus experiencing a mothering role through a relaxed context such as play might prove beneficial. A lack of such practice has been associated with an increased likelihood of females becoming aggressive mothers (Lancaster, 1971). Thus play as social practice might be adaptive. However, it is also argued that play does not represent a developmental stage of the equivalent adult behavior (Poole, 1985), because, for one, adults in some species demonstrate play (Hall, 1998).

THE FUTURE OF PLAY RESEARCH

Great ape social behavior remains of significant interest to primatologists and ethologists alike, and the study of social play has been a part of that interest for many workers. However, in comparison with other behaviors whose functions are more easily identified, play behavior has been relatively ignored. Recent research is re-identifying the importance of social play behavior on developmental, neurological, and evolutionary levels. Given the consensus that great apes exhibit sociocognitive abilities beyond that of other nonhominoid primates (e.g., incipient theory of mind, see Byrne & Whiten, 1988), it would be interesting if future work were to test for differences in those play behaviors that are shared with humans.

REFERENCES

Allen, C., & Bekoff, M. (1997). *Species of mind: The philosophy and biology of cognitive ethology.* Cambridge, MA: MIT Press.

Andrew, R. J. (1963). The origin and evolution of the calls and facial expressions of the primates. *Behaviour, 20,* 1–109.

Baldwin, J. D., & Baldwin, J. I. (1973). The role of play in social organization: Comparative observations on squirrel monkeys (*Saimiri*). *Primates, 14,* 369–381.

Barton, R. A. (2002). How did brains evolve? *Nature, 415,* 134–135.

Barton, R. A., & Harvey, P. H. (2000). Mosaic evolution of brain structure in mammals. *Nature, 405,* 1055–1058.

Bekoff, M. (1974). Social play and play-soliciting by infant canids. *American Zoologist, 14,* 323–340.

Bekoff, M. (1975). The communication of play intention: Are play signals functional? *Semiotica, 15,* 231–239.

Bekoff, M. (1977). Social communication in canids: Evidence for the evolution of a stereotyped mammalian display. *Science, 197,* 1097–1099.

Bekoff, M. (1978). Social play: Structure, function, and the evolution of a cooperative social behavior. In G. M. Burghardt & M. Bekoff (Eds.), *The development of behavior* (pp. 367–383). New York: Garland.

Bekoff, M. (2001a). The evolution of animal play, emotions, and social morality: On science, theology, spirituality, personhood, and love. *Zygon, 36,* 615–655.

Bekoff, M. (2001b). Social play behavior: Cooperation, fairness, trust, and the evolution of morality. *Journal of Consciousness Studies, 8*, 81–90.

Bekoff, M., & Allen, C. (1998). Intentional communication and social play: How and why animals negotiate and agree to play. In M. Bekoff & J. A. Byers (Eds.), *Animal play: Evolutionary, comparative and ecological perspectives* (pp. 97–114). Cambridge, UK: Cambridge University Press.

Bekoff, M., & Byers, J. A. (1981). A critical re-analysis of the ontogeny and phylogeny of mammalian social and locomotor play: An ethological hornet's nest. In K. Immelmann, G. Barlow, M. Main, & L. Petrinovich (Eds.), *Behavioral development* (pp. 296–337). Cambridge, UK: Cambridge University Press.

Bekoff, M., & Byers, J. A. (1985). The development of behavior from evolutionary and ecological perspectives in mammals and birds. *Evolutionary Biology, 19*, 215–286.

Bekoff, M., & Byers, J. A. (Eds.). (1998). *Animal play: Evolutionary, comparative, and ecological perspectives.* Cambridge, UK: Cambridge University Press.

Berman, C. M. (1983). Early differences in relationships between infants and other group members based on the mother's status: their possible relationships to peer–peer acquisition. In R. A. Hinde (Ed.), *Primate social relationships: An integrated approach* (pp. 154–156). Oxford, UK: Blackwell.

Bloomsmith, M. A. (1989). Interactions between adult male and immature captive chimpanzees: Implications for housing chimpanzees. *American Journal of Primatology* (Suppl. 1), 93–99.

Bloomsmith, M. A., Pazol, K. A., & Alford, P. L. (1994). Juvenile and adolescent chimpanzee behavioral development in complex groups. *Applied Animal Behaviour Science, 39*, 73–87.

Bramblett, C. A. (1976). *Patterns of primate play behavior.* Palo Alto, CA: Mayfield.

Brown, S. G. (1988). Play behavior in lowland gorillas: Age-differences, sex-differences, and possible functions. *Primates, 29*, 219–228.

Burghardt, G. M. (1984). On the origins of play. In P. K. Smith (Ed.), *Play in animals and humans* (pp. 5–41). New York: Blackwell.

Burghardt, G. M. (2001). Play: Attributes and neural substrates. In E. Blass (Ed.), *Developmental psychobiology: Vol. 13. Handbook of behavioral neurobiology* (pp. 327–366). New York: Kluwer Academic/Plenum Press.

Byers, J. A. (1984). Play in ungulates. In P. K. Smith (Ed.), *Play in animals and humans* (pp. 43–69). New York: Blackwell.

Byers, J. A. & Walker, C. (1995). Refining the motor training hypothesis for the evolution of play. *American Naturalist, 146*, 25–40.

Byrne, R. W., Conning, A. M., & Young, J. (1983). Social relationships in a captive group of diana monkeys (*Cercopithecus diana*). *Primates, 24*, 360–370.

Byrne, R. W., & Whiten, A. (Eds.) (1988). *Machiavellian intelligence: Social expertise and the evolution of intellect in monkeys, apes, and humans.* Oxford, UK: Clarendon Press.

Casey, B. J., Giedd, J. N., & Thomas, K. M. (2000). Structural and functional brain development and its relation to cognitive development. *Biological Psychology, 54*, 241–257.

Chalmers, N. R. (1984). Social play in monkeys: Theories and data. In P. K. Smith (Ed.), *Play in animals and humans* (pp. 119–141). New York: Blackwell.

Chiszar, D. (1985). Ontogeny of communicative behaviors. In E. S. Gollin (Ed.), *The comparative development of adaptive skills: Evolutionary implications* (pp. 207–238), Hillsdale, NJ: Erlbaum.

Clark, D. A., Mitra, P. P., & Wang, S. S. H. (2001). Scalable architecture in mammalian brains. *Nature, 411*, 189–193.

Colvin, J. (1983). Rank influences rhesus male peer relationships. In R. A. Hinde (Ed.), *Primate social relationships: An integrated approach* (pp. 57–64). Oxford, UK: Blackwell.

Dąmbska, M., & Kuchna, I. (1996). Different developmental rates of selected brain structures in humans. *Acta Neurobiologiæ Experimentalis, 56*, 83–93.

Darwin, C. R. (1872). *The expression of the emotions in man and animals.* New York: Appleton.

deWaal, F. B. M. (1995). Bonobo sex and society: The behavior of a close relative challenges assumptions about male supremacy in human evolution. *Scientific American, 272*, 82–88.

deWaal, F. B. M. (1996). *Good natured: The origins of right and wrong in humans and in other animals.* Cambridge, MA: Harvard University Press.

deWaal, F. B. M., & van Hooff, J. A. R. A. M. (1981). Side-directed communication and agonistic interactions in chimpanzees. *Behaviour, 77*, 164–198.

Diamond, A. (2000). Close interrelation of motor development and cognitive development and of the cerebellum and prefrontal cortex. *Child Development, 71*, 44–56.

Dolhinow, P. J. (1971). At play in the fields. *Natural History, 80*, 66–61.

Dunbar, R. I. M. (1992). Neocortex size as a constraint on group size in primates. *Journal of Human Evolution, 20*, 469–493.

Dunbar, R. I. M. (1995). Neocortex size and group size in primates: A test of the hypothesis. *Journal of Human Evolution, 28*, 287–296.

Dunbar, R. I. M., & Bever, J. (1998). Neocortex size predicts group size in carnivores and some insectivores. *Ethology, 104*, 695–708.

Eaton, G. G., Johnson, D. F., Glick, B. B., & Worlein, J. M. (1986). Japanese macaque' (*Macaca fuscata*) social development: Sex differences in juvenile behavior. *Primates, 27*, 141–150.

Enomoto, T. (1990). Social play and sexual behavior of the bonobo (*Pan paniscus*) with special reference to flexibility. *Primates, 31*, 469–480.

Fagen, R. M. (1981). *Animal play behavior.* New York: Oxford University Press.

Fagen, R. (1992). Play, fun and the communication of well-being. *Play and Culture, 5*, 40–58.

Fagen, R. M. (1993). Primate juveniles and primate play. In M. E. Pereira & L. A. Fairbanks (Eds.), *Juvenile primates: Life history, development, and behavior* (pp. 182–196). Oxford, UK: Oxford University Press.

Fairbanks, L. A. (2000). The developmental timing of primate play: A neural selection model. In S. T. Parker, J. Langer, & M. L. McKinney (Eds.), *Biology, brains, and behavior: The evolution of human development* (pp. 131–158). Santa Fe, NM: SAR Press.

Fischer, R. B., & Nadler, R. D. (1978). Affiliative, playful, and homosexual interactions of adult female lowland gorillas. *Primates, 19*, 657–664.

Förderreuther, B., & Zimmermann, E. (2003, October). *"Laughter" in bonobos? Preliminary results of an acoustic analysis of tickling sounds in a hand-reared male.* Paper presented at the annual meeting of the German Primate Society, Leipzig, Germany.

Fossey, D. (1979). Development of the mountain gorilla (*Gorilla gorilla beringei*): The first thirty-six months. In D. A. Hamburg & E. R. McCown (Eds.), *The great apes* (pp. 139–184). Menlo Park, CA: Benjamin/Cummings.

Freeman, H. E., & Alcock, J. (1973). Play behavior of a mixed group of juvenile gorillas and orang-utans, *Gorilla g. gorilla* and *Pongo p. pygmaeus. International Zoo Yearbook, 13*, 189–194.

Fruth, B., & Hohmann, G. (1996). Comparative analyses of nest building behavior in bonobos and chimpanzees. In R. W. Wrangham, W. C. McGrew, F. M. B. deWaal, & P. G. Heltne (Eds.), *Chimpanzee cultures* (pp. 109–128). Cambridge, MA: Harvard University Press.

Govindarajulu, P., Hunte, W., Vermeer, L. A., & Horrocks, J. A. (1993). The ontogeny of social play in a feral troop of vervet monkeys (*Cercopithecus aethiops sabaeus*): The function of early play. *International Journal of Primatology, 14*, 701–719.

Hall, S. L. (1998). Object play by adult animals. In M. Bekoff & J. A. Byers (Eds.), *Animal play: Evolutionary, comparative and ecological perspectives* (pp. 45–60). Cambridge, UK: Cambridge University Press.

Harrison, B. (1961). A study of orang-utan behavior in the semi-wild state. *International Zoo Yearbook, 3*, 57–68.

Harrison, B. (1962). *Orang-utan*. New York: Doubleday.

Hashimoto, C. (1997). Context and development of sexual behavior of wild bonobos (*Pan paniscus*) at Wamba, Zaire. *International Journal of Primatology, 18*, 1–21.

Hashimoto, C., & Furuichi, T. (1996). Social role and development of noncopulatory sexual behavior of wild bonobos. In R. W. Wrangham, W. C. McGrew, F. M. B. deWaal, & P. G. Heltne (Eds.), *Chimpanzee cultures* (pp. 155–168). Cambridge, MA: Harvard University Press.

Horvat, J. R., & Kraemer, H. C. (1981). Infant socialization and maternal influence in chimpanzees. *Folia Primatologica, 36*, 99–110.

Hrdy, S. B. (1976). Care and exploitation of nonhuman primate infants by conspecifics other than the mother. In J. S. Rosenblatt, R. A. Hinde, E. Shaw, & C. Beer (Eds.) *Advances in the study of behavior* (Vol. 6, pp. 101–158). New York: Academic Press.

Humphreys, A. P., & Smith, P. K. (1987). Rough and tumble, friendship, and dominance in school children: Evidence for continuity and change with age. *Child Development, 58*, 201–212.

Iwaniuk, A. N., Nelson, J. E., & Pellis, S. M. (2001). Do big-brained animals play more? Comparative analyses of play and relative brain size in mammals. *Journal of Comparative Psychology, 115*, 29–41.

Janson, C. H., & Van Schaik, C. P. (1993). Ecological risk aversion in juvenile primates: Slow and steady wins the race. In M. E. Pereira & L. A. Fairbanks (Eds.), *Juvenile primates: Life history, development, and behavior* (pp. 57–76). Oxford, UK: Oxford University Press.

Joffe, T. H. (1997). Social pressures have selected for an extended juvenile period in primates. *Journal of Human Evolution, 32*, 593–605.

Jolly, A. (1985). *The evolution of primate behavior* (2nd ed.). New York: Macmillan.

Joseph, R. (2000). Fetal brain behavior and cognitive development. *Developmental Review, 20*, 81–98.

King, N. E., Stevens, V. J., & Mellen, J. D. (1980). Social behavior in a captive chimpanzee (*Pan troglodytes*) group. *Primates, 21*, 198–210.

Kipper, S., & Todt, D. (2001). Variation of sound parameters affects the evaluation of human laughter. *Behavior, 138*, 1161–1178.

Koyama, N. (1985). Playmate relationships among individuals of the Japanese monkey troop in Arashiyama. *Primates, 26*, 390–406.

Kraemar, H. C., Horvat, J. R., Doering, C., & McGinnis, P. R. (1982). Male chimpanzee development focusing on adolescence: Integration of behavioral with physical changes. *Primates, 23*, 393–405.

Lancaster, J. B. (1971). Play-mothering: The relations between juvenile females and young infants among free-ranging vervet monkeys (*Cercopithecus aethiops*). *Folia Primatologica, 15*, 161–182.

Lee, P. C. (1983). Play as a means for developing relationships. In R. A. Hinde (Ed.), *Primate social relationships: An integrated approach* (pp. 82–89). Oxford, UK: Blackwell.

Lewis, K. P. (2000). A comparative study of primate play behaviour: Implications for the study of cognition. *Folia Primatologica, 71*, 417–421.

Lewis, K. P. (2003). *A comparative analysis of play behaviour in primates and carnivores.* Unpublished doctoral dissertation, University of Durham, Durham, UK.

Lewis, K. P., & Barton, R. A. (2004). Playing for keeps: Evolutionary relationships between the cerebellum and social play behaviour in non-human primates. *Human Nature, 15*, 5–22.

Lewis, K. P., & Barton, R. A. (2004). *Amygdala size and hypothalamus size predict social play frequency in non-human primates: A comparative analysis using independent contrasts.* Manuscript submitted for publication.

Markus, N., & Croft, D. B. (1995). Play behavior and its effects on social development of common chimpanzees (*Pan troglodytes*). *Primates, 36*, 213–225.

Marler, P., & Tenaza, R. (1977). Signalling behavior of apes with special reference to vocalization. In T. A. Seboek (Ed.), *How animals communicate* (pp. 965–1033). Bloomington: Indiana University Press.

Martin, P., & Caro, T. M. (1985). On the functions of play and its role in behavioural development. In J. Rosenblatt, C. Beer, M. Busnel, & P. Slater, P.

(Eds.), *Advances in the study of behaviour* (Vol. 15, pp. 59–103). New York: Academic Press.

Masataka, N., & Kohda, M. (1988). Primate play vocalizations and their functional significance. *Folia Primatologica, 50*, 152–156.

McGrew, W. C. (1996). Tools compared. In R. W. Wrangham, W. C. McGrew, F. M. B. deWaal, & P. G. Heltne (Eds.), *Chimpanzee cultures* (pp. 25–40). Cambridge, MA: Harvard University Press.

Meaney, M. J., Stewart, J., & Beatty, W. W. (1985). Sex differences in social play: The socialization of sex roles. In J. Rosenblatt, C. Beer, M. Busnel, & P. Slater (Eds.), *Advances in the study of behaviour* (Vol. 15, pp. 1–58). New York: Academic Press.

Meder, A. (1990). Sex differences in the behavior of immature captive lowland gorillas. *Primates, 31*, 51–63.

Mendoza-Granados, D., & Sommer, V. (1995). Play in chimpanzees of the Arnhem Zoo: Self-serving compromises. *Primates, 36*, 57–68.

Merrick, N. J. (1977). Social grooming and play behavior of a captive group of chimpanzees. *Primates, 18*, 215–224.

Nash, L. T. (1993). Juveniles in nongregarious primates. In M. E. Pereira & L. A. Fairbanks (Eds.), *Juvenile primates: Life history, development, and behavior* (pp. 119–137). Oxford, UK: Oxford University Press.

Owens, N. W. (1975). A comparison of aggressive play and aggression in free-living baboons, *Papio anubis. Animal Behaviour, 23*, 757–765.

Panksepp, J. (1998). *Affective neuroscience: The foundations of human and animal emotions.* New York: Oxford University Press.

Panksepp, J., & Burgdorf, J. (2000). 50-khz chirping (laughter?) in response to conditioned and unconditioned tickle-response reward in rats: Effects of social housing and genetic variables. *Behavioral Brain Research, 115*, 25–38.

Panksepp, J., Normansell, L., Cox, J. F., & Siviy, S. M. (1994). Effects of neonatal decortication on the social play of juvenile rats. *Physiology and Behavior, 56*, 429–443.

Paquette, D. (1994). Fighting and play-fighting in captive adolescent chimpanzees. *Aggressive Behavior, 20*, 49–65.

Pellegrini, A. D. (1995). Boys' rough-and-tumble play and social competence: Contemporaneous and longitudinal relations. In A. D. Pellegrini (Ed.), *The future of play theory: A multidisciplinary inquiry into the contributions of Brian Sutton-Smith* (pp. 107–126). Albany: State University of New York Press.

Pellegrini, A. D., & Smith, P. K. (1998). The development of play during childhood: Forms and possible functions. *Child Psychology and Psychiatry Review, 3*, 51–57.

Pellis, S. M., & Pellis, V. C. (1996). On knowing it's only play: The role of play signals in play-fighting. *Aggressive Behavior, 1*, 249–268.

Pereira, M. E., & Preisser, M. C. (1998). Do strong primate players "self-handicap" during competitive social play? *Folia Primatologica, 69*, 177–180.

Poole, T. B. (1985). *Social behaviour in mammals.* London: Blackie.

Poole, T. B., & Fish, J. (1975). An investigation of playful behaviour in *Rattus norvegicus* and *Mus musculus* (*Mammalia*). *Journal of Zoology, 175*, 61–71.

Pruetz, J. D., & Bloomsmith, M. A. (1995). The effects of paternity on interactions between adult male and immature chimpanzees in captivity. *Folia Primatologica, 65*, 174–180.

Pusey, A. E. (1990). Behavioral changes at adolescence in chimpanzees. *Behavior, 115*, 203–246.

Quartz, S. R., & Sejnowski, T. J. (1997). The neural basis of cognitive development: A constructivist manifesto. *Behavioral and Brain Sciences, 20*, 537.

Ramachandran, V. S. (1998). The neurology and evolution of humor, laughter, and smiling: The false alarm theory. *Medical Hypotheses, 51*, 351–354.

Rijksen, H. D. (1978). *A field study on Sumatran Orang-utans.* Wageningen, Netherlands: Veenman & Zonen.

Rowe, N. (1996). *The pictorial guide to the living primates.* New York: Pogonias.

Rowell, T. E., & Chism, J. (1986). The ontogeny of sex differences in the behavior of Patas monkeys. *International Journal of Primatology, 7*, 83–105.

Ruppenthal, G. C., Arling, G. L, Harlow, H. F., Sackett, G. P., & Suomi, S. J. (1976). A 10-year perspective of motherless–mother monkey behavior. *Journal of Abnormal Psychology, 85*, 341–349.

Schåfer, M., & Smith, P. K. (1996). Teachers' perceptions of play fighting and real fighting in primary school. *Educational Research, 38*, 173–181.

Schaller, G. B. (1972). The behavior of the mountain gorilla. In P. Dolhinow (Ed.), *Primate patterns* (pp. 85–124). New York: Holt, Reinhart & Winston.

Siviy, S. M., & Panksepp, J. (1987). Juvenile play in the rat: Thalamic and brain stem involvement. *Physiology and Behavior, 41*, 103–114.

Smith, P. K. (1982). Does play matter? Functional and evolutionary aspects of animal and human play. *Behavioral and Brain Sciences, 5*, 139–184.

Spijkerman, R. P., van Hooff, J. A. R. A. M., Dienske, H., & Jens, W. (1997). Differences in subadult behaviors of chimpanzees living in peer groups and in a family group. *International Journal of Primatology, 18*, 439–454.

Špinka, M., Newberry, R. C., & Bekoff, M. (2001). Mammalian play: Training for the unexpected. *Quarterly Review of Biology, 76*, 141–168.

Suomi, S., & Harlow, H. (1972). Social rehabilitation of isolate-reared monkeys. *Developmental Psychology, 6*, 487–496.

Symons, D. (1978). *Play and aggression: A study of rhesus monkeys.* New York: Columbia University Press.

Tanner, J. E., & Byrne, R. W. (1993). Concealing facial evidence of mood: Perspective-taking in a captive gorilla. *Primates, 34*, 451–457.

Tomasello, M., Gust, D. A., & Evans, A. (1990). Peer interaction in infant chimpanzees. *Folia Primatologica, 55*, 33–40.

Vankova, D., & Bartos, L. (2002). The function of mounting behaviour in farmed red deer calves. *Ethology, 108*, 473–482.

van Lawick-Goodall, J. (1968). The behaviour of free-living chimpanzees in the Gombe Stream Reserve. *Animal Behaviour Monographs, 1*, 161–311.

Van Schaik, C. P., Ancrenaz, M., Borgen, G., Galdikas, B., Knott, C. D., Singleton, I., Suzuki, A., Utami, S. S., & Merrill, M. (2003). Orangutan cultures and the evolution of material culture. *Science, 299*, 102–105.

Voland, E. (1977). Social play behavior of the common marmoset (*Callithrix jacchus* Erxl., 1777) in captivity. *Primates, 18*, 883–901.

Watson, D. M. (1993). The play associations of red-necked wallabies (*Macropus rufogriseus banksianus*) and relation to other social contexts. *Ethology, 94*, 1–20.

Watson, D. M. (1998). Kangaroos at play: Play behaviour in the *Macropodoidea*. In M. Bekoff & J. A. Byers (Eds.), *Animal play: Evolutionary, comparative, and ecological perspectives* (pp. 61–95). Cambridge, UK: Cambridge University Press.

Whiten, A., & Byrne, R. W. (Eds.). (1997). *Machiavellian intelligence II: Extensions and evaluations.* Cambridge, UK: Cambridge University Press.

Whiting, B. A. (2002). *The evolution of the primate cerebellum.* Unpublished master's thesis, University of Durham, Durham, UK.

Willingham, D. B. (1999). The neural basis of motor-skill learning. *Current Directions in Psychological Science, 6*, 178–182

Zahavi, A. (1977). The testing of a bond. *Animal Behaviour, 25*, 246–247.

Zucker, E. L., Mitchell, G., & Maple, T. (1978). Adult male-offspring play interactions within a captive group of orang-utans (*Pongo pygmaeus*). *Primates, 19*, 379–384.

CHAPTER 4

Rough-and-Tumble Social Play in Humans

DOUGLAS P. FRY

A great deal of play in humans involves interaction with others in the social world. Social play is diverse, involving physical games, running, jumping, and wrestling as well as a plethora of jointly created make-believe enactments of social scenes, activities, and rituals. Two or more children may assume particular roles during social play, such as "mother" and "children," "hunters" and "prey," "buyers" and "sellers," or "wedding guests," "bride," "groom," and "priest." As we will see later, the imitation of adult activities forms a prominent feature of social play across cultural settings.

This chapter focuses on one form of social play called rough-and-tumble (R&T) in humans. The term "rough-and-tumble play" was originally used to describe play chasing, fleeing, and wrestling in rhesus monkeys (Harlow & Harlow, 1965), but R&T also occurs in many other animal species (Aldis, 1975; Einon & Potegal, 1991; Fagen, 1981; Pellis, Field, Smith, & Pellis, 1996; Smith, 1982). In primates, Jolly (1985, p. 406) notes, "chasing and wresting with peers is ubiquitous. Every species has rough-and-tumble." A number of features typify R&T in mammals: Threats are absent or infrequent, movements are free and easy, muscle tone is relaxed, biting is inhibited, play signals such as the play face and play vocalizations are evident, roles frequently reverse, dominance relations are relaxed, animals of different sizes are partners, and sequences of behavior vary (Aldis, 1975; Bekoff & Byers, 1981, 1985; Fagen, 1978, 1981; Pellis, 1984; Smith, 1982; Symons, 1978).

It seems very unlikely that R&T would be widespread among mammals, including primates, if it did not contribute to individual survival and reproductive success. Based on a detailed study of R&T in rhesus monkeys, Symons (1978, p. 196) concluded that "the practice and perfection of skills in predator avoidance and aggression is an adaptive function of aggressive play, in the rigorous sense that aggressive play was shaped by natural selection for these purposes."

The observations that the overall pattern of R&T in children is widely, probably universally, distributed across human cultures and that a tendency exists for boys to engage in more R&T and more vigorous R&T than girls, coupled with research findings on the nature of R&T across children of different ages suggest that, as in other species, play fighting in humans has evolved to fulfill certain adaptive functions (Boulton, 1996; Boulton & Smith, 1992; Fry, 1990; Pellegrini, 1994, 2002; Smith, 1997; Smith & Boulton, 1990). At the same time, a cross-cultural examination of play fighting reveals that stylistic variations are clearly learned in different human social environments. Social and physical environmental factors also influence the frequency of play fighting (Fry, 1992). Human play fighting provides an interesting example of how evolutionary factors *and* elements of the social and physical environment interact.

The observation that cross-cultural similarities and differences exist in R&T suggests that a musical metaphor, "variations on a theme," aptly applies. In certain musical works, ranging from symphonic to jazz compositions, a musical motif—a theme—recurs in alternate but recognizable forms—the variations—perhaps at different tempos or played by different instruments. Regarding R&T, the "theme" is that children everywhere engage in recognizable play fighting and play chasing, and as a generalization, boys engage in more R&T than do girls. "Variations" come into play, for instance, as culturally specific elements are incorporated into R&T and as cultural attitudes that reflect different tolerance levels for R&T influence the behavior. The variations on a theme metaphor may serve as a useful reminder that both cross-cultural patterns and culturally based variations exist simultaneously.

The preliminary evidence suggests that the practice of fighting skills and establishment of dominance are likely evolutionary functions that shift in importance across the age span of the individuals engaging in R&T

(Boulton & Smith, 1992; Pellegrini, 1994, 2002; Smith, 1982, 1997). Specifically, the importance of a practice function in young children may be supplanted, to some degree, by a dominance function in later childhood (Pellegrini, 1994, 2002; Smith & Boulton, 1990). Additionally, the anthropological data hint at the possibility that the R&T of older children may have some intriguing parallels with the competitive contests of adults. The exploration of functional explanations opens the door to a brief discussion of human aggression and contests in relation to R&T. Contests are discussed as culturally formalized events that have some features of both play and aggression—hence, paralleling the R&T bouts of adolescents.

FEATURES OF R&T IN HUMANS

Smith (1997, p. 48) defines R&T as "a cluster of behaviours whose core is rough but playful wrestling and tumbling on the ground; and whose general characteristic is that the behaviours seem to be agonistic but in a non-serious, playful context." The majority of detailed studies dealing with children's R&T have been conducted in the United Kingdom (e.g., Blurton Jones, 1967, 1972; Boulton, 1991a, 1991b, 1993a, 1993b, 1996; Humphreys & Smith, 1984, 1987; McGrew, 1972; Nabuzoka & Smith, 1999; Smith & Boulton, 1990; Smith & Lewis, 1985) and in the United States (e.g., Aldis, 1975; Pellegrini, 1988, 1989, 1993, 1994, 1995, 2002, 2003; Scott & Panksepp, 2003). Human R&T consists of chasing, fleeing, wrestling, grappling, and delivering restrained blows, usually with laughter or other clear indicators of playful intent (Blurton Jones, 1972; Fry, 1990; Smith & Lewis, 1985). For example, Humphreys and Smith (1987) specify R&T to have subcategories of grappling, flailing, and chasing among 7-, 9-, and 11-year-old British children. *Grappling* consists of the specific behavioral elements of hand wrestling, wrestling while standing, wrestling while lying, holding, lying on, sitting on, pushing and pulling. *Flailing* includes hitting at and hitting, kicking at and kicking. *Chasing* refers to chasing and fleeing. Pellegrini (1988, 1989; see also Scott & Panksepp, 2003) was able to use these same behavioral categories during his study of 5- to 10-year-olds in the southern United States—a fact that suggests a high degree of correspondence in R&T between these two national settings.

Aldis (1975) focused on play fighting in 6- to 12-year-olds in California and coined terms for two types of wrestling. First, "wrestling for superior position" occurs almost exclusively among boys. Aldis explains that "one boy will make strenuous efforts to throw another to the ground, to get on top of him, to hold him down, to flatten him, and sometimes to pin him to the ground. Pinning constitutes a symbolic play 'victory'" (p. 178). The second pattern, called "fragmentary wrestling," is less vigorous and involves both girls and boys: "neither child will attempt to achieve a clear-cut superiority but will merely grapple or push and pull in various directions" (p. 178). Aldis observes that in the presence of laughter, hitting, kicking, chasing, and fleeing co-occur with these types of wrestling.

DISTINCTIONS BETWEEN R&T AND AGGRESSION

R&T and real fighting, although superficially similar to each other, can be distinguished (Blurton Jones, 1967; Boulton, 1994, 1996; Fry, 1987; Pellegrini, 2002; Smith, 1997; Smith & Lewis, 1985). After surveying the literature on play fighting in various animal species, Bekoff (1981) and Smith (1982) pointed out that fighting and R&T are structurally different. Thus a phylogenetic precedent exists for distinguishing between R&T and aggression.

Descriptions of play aggression usually mention laughs, smiles, or play faces as signals of friendly relations among the participants, whether the descriptions are provided by researchers who mention the phenomenon in passing (e.g., Eibl-Eibesfeldt, 1974) or by authors who describe R&T in greater detail (e.g., Blurton Jones, 1967). By contrast, real fighting differs from play fighting in terms of expressions, gestures, and postures: low frowns, bared teeth, leaning forward, clenched fists, fixated gazes and crying with a puckered facial expression (Blurton Jones, 1967, 1972; Camras, 1980; Eibl-Eibesfeldt, 1974; Fry, 1987). Among children of 8 years or less, play and aggressive signals rarely co-occur during the same interaction (Fry, 1987; Smith & Lewis, 1985). Among Zapotec children of southern Mexico, Fry (1987) observed that occasionally play fighting occurred with a neutral facial expressions. Contextual cues and the absence of aggressive signals suggested playful intent.

Several studies have found that children tend to remain in each other's company after R&T but are more likely to separate following aggression (Aldis, 1975; Fry, 1990; Humphreys & Smith, 1984; Smith & Lewis, 1985). Using facial expressions in conjunction with spatial cues, Smith and Lewis (1985) were able to classify the vast majority of events (about 97%) in British 3- to 4-year-olds as either play or aggression. Furthermore, children themselves can differentiate fighting from play fighting (Costabile et al.,

1991; Smith, Hunter, Carvalho, & Costabile, 1992; Smith & Lewis, 1985; Smith, Smees, & Pellegrini, 2004).

After observing monkeys, Bateson (1972) proposed that play signals function in a metacommunicative manner to connote, for example, that a nip is playful, not real. Symons (1978, p. 96) proposes, however, that such metacommunication is actually not necessary for explaining play signals: "When a monkey nips playfully it probably means precisely that, and when it emits a specialized signal before or during playfighting, that signal may refer to the playfighting, and may function to reduce the probability of aggression, without contrasting playfighting with nonexistent or fictional entities (e.g., a real bite)." Symons's point is that whereas a contrast between play and aggression may have utility for human observers, it may be superfluous for monkeys.

Turning to humans, children do utilize a concept of play in contrast to aggression, as indicated by the verbalizations of playful intent they direct to their peers (Aldis, 1975; Smith & Lewis, 1985). Therefore, Bateson's idea of metacommunication does apply to human children. We consider further intricacies of metacommunication in humans, regarding play vis-à-vis aggression, at the end of the chapter.

Fry (1987) draws on the animal R&T literature (Bekoff & Byers, 1981, 1985; Fagen, 1981) and ethological observations of Zapotec children to compile a list of eight contrasts between play fighting and fighting. Smith (1997) and Smith, Smees, Pellegrini, and Menesini (2002) have updated and expanded the contrasts, as summarized in Table 4.1.

Play fighting is much more varied than fighting and sometimes clearly incorporates elements of make-believe, or fantasy (Pellegrini, 2002, p. 440). Among Zapotec children, for instance, less variable aggression usually consists of punches, pushes, kicks, and beats, but only rarely more "exotic," that is, fantasy-entailing attacks like roundhouse kicks or karate chops, as sometimes occur during R&T bouts (Fry, 1987, p. 293). British children engage in boxing and karate, use pretend guns, and adopt roles such as "police" and "prisoner" during their R&T (Humphreys & Smith, 1984; Smith et al., 2004). Hunter-gatherer Zhun/twa children of the African Kalahari Desert sometimes incorporate an animal hunt into their R&T, as explained by Konner (1972, p. 300): "Rough and tumble play is usually either stylized, as when one child pretends to be an animal that the others are attacking, or else it takes a mild form that consists, in spite of the available space, of laughing, hugging, and rolling around on the ground together." Along with pretend elements, activities such as running, jumping, twirling around, and crouching down often are interspersed with R&T but rarely with real fighting (Fry, 1987).

TABLE 4.1. Distinguishing Features of R&T and Aggression

Feature	Play	Aggression
1. Facial expression	Smile, laugh, play face, or, less often, neutral face	Frown, fixated stare, bared teeth, crying, puckered face
2. Conflict over a resource precedes the encounter	Rare	Frequent
3. Variable behavioral content during an encounter	Typical	Rare
4. Fantasy elements	Common	Absent
5. Number of participants	Two or more	Rarely more than two
6. Role reversals	Common	Absent
7. Self-handicapping (by larger/stronger partner)	Occurs	Absent
8. Restraint	Typical	Less apparent or absent
9. Chase–flee	Sometimes	Rare
10. Reactions of others	Typically not interested	Encounter draws a crowd
11. Wrestling	Sometimes (with restraint)	Rare (but forceful)
12. Relative frequency	Many times more common than aggression	Many times less common than play
13. Spatial relation of participants following the encounter	Often remain together	Often separate

Note. See Fry (1987, Table 3), Smith (1997, Table 1), and Smith et al. (2002, Table 14.1).

R&T AS FUN, FRIENDLY, AND AFFILIATIVE

When asked why they play fight, children typically answer, because "it's fun" (Smith & Boulton, 1990, p. 227; see also Humphreys & Smith, 1987; Pellegrini, 2002; Smith & Lewis, 1985; Smith et al., 2002, 2004). Most R&T researchers concur that play fighting is a social, affiliative activity (Blurton Jones, 1972; Fry, 1987, 1990; Scott & Panksepp, 2003; Smith & Boulton, 1990), although there are exceptions involving socially rejected children (Pellegrini, 1988, 1993, 1995). On the basis of interviews with children, Smith and colleagues (2002; see also Smith et al., 2004) report that the vast majority (85%) of play-fighting partners are friends, and half of these partners are considered best friends. The social affiliative view of play fighting also is reflected in Whiting and Whiting's (1975) use of the term "assaults sociably" for R&T in their study of six cultures.

Studies of R&T in school playground settings show it to occupy between 5 and 10% of free playtime, which is much more frequent than real fighting (generally, less than 1% of playground time) (Humphreys & Smith, 1984; Neill, 1976; Pellegrini, 1989; Smith, 1997; Smith & Boulton, 1990; Smith & Lewis, 1985). Correspondingly, play fighting occurs nine times more often than aggression among Zapotec children (Fry, 1987). Humphreys and Smith (1987) report that the majority of 7-, 9-, and 11-year-old British children in their sample stayed together following R&T. Similarly, Zapotec children were significantly more likely to stay together after play fighting than following an aggressive interaction (Fry, 1990). The frequent occurrence of R&T and the greater likelihood of staying together after R&T than after real aggression correspond with the inference that play fighting is socially positive, friendly, and fun, as contrasted to fighting. Furthermore, Humphreys and Smith point out that R&T can be viewed as reflecting a "mutual wish" by participants to engage in this activity. In their study, about one-third of play initiations were accepted. Clearly, play initiations are not always accepted by the recipient. A similar percentage of play-fighting acceptances (31%) was observed among Zapotec children (Fry, 1990). Aggressive acts, however, yielded a response only 8% of the time among the Zapotec children. This difference between play and aggression provides another indication that play fighting is a socially positive activity, in contrast to fighting.

R&T and real aggression thus seem to reflect antithetical moods in young children (Fry, 1987, 1990). Play is enjoyable, and children engage in it much more frequently and for longer periods of time than they engage in fighting. Children are more likely to remain together following play than after aggression. Children also are more likely to respond reciprocally to playful acts than aggressive acts. Additionally, R&T is significantly more likely to involve more than two participants than is fighting, reflecting a willingness, or even an eagerness, to join others in the activity (Fry, 1987; Smith, 1997). In short, children say that R&T is fun, and they laugh and smile while engaging in it.

CUES THAT DISTINGUISH PLAY FROM AGGRESSION

Several studies have found high levels of consensus among viewers as to whether particular interactions are real aggression or play fighting (Boulton, 1993b; Costabile et al., 1991; Schåfer & Smith, 1996; Smith & Lewis, 1985). For instance, Costabile and colleagues (1991) showed videotapes of interacting children from Italy and England to 8- and 11-year-old children from both countries. Following each taped episode, independently viewing children were asked if either playful or real fighting was portrayed, and how did they know? The 11-year-old children performed the task slightly but significantly better than the 8-year-olds (with 87% and 84% accuracy, respectively). The results were similar for children from both countries, and it did not matter whether children were viewing episodes filmed in their own or in the other country. Most frequently, the children said that they made their distinctions between play and aggression based on physical actions (e.g., "It was only a play fight 'cos he didn't hit him hard," or "That was a real fight because he ran after him and kicked him in the leg") and inferences about affect (e.g., "It was a real fight because they were both angry") or inferences about intent (e.g., It was a true fight. . . . He tried his best to get him") (Smith, 1997, p. 52).

In another study, Smith and colleagues (1992) interviewed Italian and English 5-, 8-, and 11-year-olds. When asked whether or not they could distinguish real fighting from play fighting, a solid majority of children across all age groups said they could do so. Among the 5-year-olds, 40% could not explain clearly how they could tell, but almost all of the children in the two older age groups could provide reasons. As in the Costabile and colleagues (1991) study, these 8- and 11-year-olds referred often to physical characteristics (i.e., actions/restraint) and inferences about affect or intent, but also stated that they relied on facial expressions and verbal cues.

Smith and colleagues (2004, p. 171; see also Pellegrini, 2003) employed an innovative methodology wherein they played back video footage of children's playful and aggressive fighting to participants and nonparticipants in the incidents. When asked whether an interaction was play fighting or real fighting and "how can you tell?", the participants in the event were able to provide more criteria for their judgments than were nonparticipants. Participants also cited reasons based on privileged information gained through their personal involvement in particular events, such as "whether a hit or kick really hurts, and whether an apparently aggressive act is within a pretend or game framework previously agreed by those involved."

UNIVERSALITY: THE R&T THEME

Smith (1997, p. 51) suggests that play fighting "may be a cultural universal, whose development is canalised to a considerable degree." To evoke the musical metaphor of variations on a theme, R&T as a "theme" recurs with "variations" across diverse cultural environments. As mentioned earlier, most of the focused research on R&T has been conducted in the United Kingdom and the United States. With only a few exceptions, data on R&T in other cultures is scant and usually not detailed. Exceptions are Konner's (1972) article on the African Zhun/twa, Frey and Hoppe-Graff's (1994) study of Brazilian 2- to 4-year-olds, and Fry's (1987, 1988, 1990, 1992) work with Zapotec 3- to 8-year-olds. Nonetheless, a recurring theme is discernible that includes the rolling and pinning of wrestling, the pushing and pulling of grappling, the reversal of roles in chasing and fleeing, and the use of restrained blows, all in the presence of play signals (see Figures 4.1 and 4.2).

Children from locations as disparate as New Guinea, Scotland, California, and Mexico attempt to get on top of and pin opponents in basically the same way, in what Aldis calls "wrestle for superior position" (Aldis, 1975;

FIGURE 4.1. Gentle grappling between a Zapotec brother and sister. The smiles signal that this interaction is playful. D. P. Fry photo collection.

FIGURE 4.2. Wrestling sequence between brothers from the United States. The boys take advantage of the dock and lake setting to repeatedly wrestle and throw each other into the water in noninjurious R&T. The left-hand photo precedes the right-hand photo by 2–3 seconds. Self-handicapping is apparent as the older brother, age 11 (facing the camera) allows his younger partner, age 7, to throw him. Note the smile. D. P. Fry photo collection.

Fry, 1987; McGrew, 1972; Sorenson, 1978). Sorenson (1978, p. 24) mentions the rough play of Fore children from New Guinea only in passing, but he does provide a sequence of photographs extracted from movie film that clearly shows two boys, of about 8 years of age, rolling on the ground as they wrestle for superior position.

As among youth in California, "fragmentary wrestling" among Zapotec children consists of behaviors such as pushes, pulls, beats, punches, slaps, and kicks, and less often, knuckle raps, backhand slaps, and karate chops. This overall pattern also mirrors elements of children's play fighting reported in other cultural contexts (Aldis, 1975; Blurton Jones, 1972; Boulton, 1996; Frey & Hoppe-Graff, 1994; McGrew, 1972; Pellegrini, 1993; Smith & Lewis, 1985). A photograph of two boys from Ganda society in Africa shows them grappling from a standing position (Roscoe, 1911, Figure 17). North American Klamath boys also wrestle from a standing position, in which "opponents face each other with hands on each other's waists" (Pearsall, 1950, p. 346-b). Gusinde writes of South American Ona boys:

> They grasp each other in such a way that the arms grip downward; each one places his two arms around the hips of the other one, locks them tightly, and folds his hands in the latter's sacral region, also moving them up to the neck during the wrestling; for greater stability the legs are spread more or less, and the upper body bent far forward. Each one now tries to lift his opponent and throw him down. (1931, p. 1624)

In their six-cultures study, Whiting and Whiting (1975, p. 61) use the behavioral labels "assaults sociably" and "horseplay" for "friendly wrestling and back-slapping." These authors note that laughter differentiates friendly assault from real aggression (Whiting & Whiting, 1975, p. 193). The Whitings report both sociable and real assaults among children in all six cultures (Japan, Philippines, Mexico, the United States, the Khalapur of India, and the Nyansongo of Kenya), but they do not describe these two patterns in much detail. Eibl-Eibesfeldt (1974) provides examples of aggression among African !Ko San hunter-gatherer children but does not describe play aggression, other than to state that play could be distinguished from real fighting by laughs and smiles. Stross (1970, p. 60) briefly describes Tzeltal Mayan R&T: "Playful wrestling among boys is as common among boys as simple chasing and falling is among girls. . . . Hopping, running, skipping, and jumping are practiced by children of both sexes."

Across cultures, laughs, play faces, and smiles appear to be universally employed and understood to denote friendly, playful intent. Children themselves state that facial expressions and vocalizations are ways to tell play from aggression (Costabile et al., 1991; Humphreys & Smith, 1987; Smith & Boulton, 1990; Smith & Lewis, 1985). The interpretation that play faces, smiles, and laughs universally signal playful intentions corresponds with Ekman and colleagues' (1987) findings that facial displays of happiness are understood panculturally. Fry (1987, p. 297) suggests that a normal or neutral facial expression may also convey a message of nonserious intent in some situations. Among Zapotec 3- to 8-year-olds, neutral facial expressions are interspersed with play signals, such as laughs, but tend not to be as closely associated with agonistic signals (Fry, 1987, p. 297).

Sex differences in R&T also are part of the cross-cultural theme. Aldis (1975) reports that boys engage in much more wrestling for superior position than do girls. Correspondingly, many studies show that boys engage more avidly in play fighting than girls (Blurton Jones & Konner, 1973; Humphreys & Smith, 1984; Pellegrini, 1987; Whiting & Edwards, 1988), although a minority of studies report no significant gender-related difference (Fry, 1988; Smith & Lewis, 1985). However, there is a paucity of findings that shows that girls engage in significantly more play fighting than boys (cf. Blurton Jones, 1972, p. 109). Pellegrini (1987, p. 29; see also Pellegrini, 2002, p. 440) sums up the situation: "one of the more reliable sex differences in the developmental literature is that boys engage in more R&T than do girls."

An anthropological database, called the Human Relations Area Files (HRAF), which has existed in paper and microfiche form for decades, is currently being expanded and transferred into electronic form. The electronic version, the eHRAF, allows users to run keyword searches of the cross-cultural database as well as search for precoded topics (e.g., "activities of children," "social control," "food preparation," and so forth). As of this writing, the eHRAF contains information on 137 cultures. Several keyword searches were run on all cultures in the eHRAF related to play fighting, R&T, and wrestling. Given the rarity of information on R&T in anthropological descriptions, searches were run on all 137 cultures in the electronic database to maximize the chance of finding relevant material. In fact, very little meaningful information resulted from keyword searches for variants of R&T and play fighting. However, combining the truncated, wild-card search term "wrestl*" with the precoded category "activities of children" resulted in "hits" for 30 cultures related to wrestling among children.

It is important to bear in mind that these findings stem from the convenience sample of all 137 societies currently in the electronic version of the eHRAF. All world continents are represented, but the number of societies in this sample varies among continents. The fact that the majority of ethnographies provide very little information on children's play should also be considered. For instance, it would be unwise to conclude, based on these search results, that children do not wrestle in the majority of the 137 societies in this sample, because missing data on children's wrestling, not its actual absence, may likely account for this situation. The fact that no ethnographer ever states that children do not wrestle also is congruent with such an interpretation.

One of the clearest patterns to emerge from this ethnographic data is a sex difference that corresponds with the previously noted R&T pattern. For the 30 societies with information on childhood wrestling, 21 descriptions link wrestling to boys, 6 to both girls and boys, and none solely to girls (sex-differentiating information was absent in the remaining 3 cases). For example, pertaining to the Pawnee of North America: "The boys are boiling over with animal spirits, and . . . are continually running about, chasing each other, wrestling, shooting arrows and playing games, of which the familiar stick game seems the favorite" (Grinnell, 1961, p. 18).

Another regularly mentioned idea to emerge as a sidelight from this search for wrestling is that children avidly imitate adult activities in their play. Imitation by children of adults was mentioned for 14 out of the 30 societies, even though imitation was *not* a direct search topic. For example, related to the Banyoro of East Africa, a Bantu-speaking people with a tradition of cattle-herding:

> There were many games, the favourites being make-believe games, in imitation of the doings of their elders. They played at going out to war, fighting battles and capturing prisoners and cattle and bringing the spoils to the king; they married and built their kraals [corrals], observing taboos as their elders did; they bought, sold and exchanged cattle and tended them, healing them of various diseases. Miniature bows and arrows were favorite toys and a kind of nine-pins or skittles, while wrestling was enjoyed by both old and young. (Roscoe, 1923, pp. 259–260)

Third, ethnographic sources for 5 of the 30 societies indicate that children sometimes incorporate fantasy into their wrestling bouts (see Aldis, 1975; Fry, 1987; Humphreys & Smith, 1984; Pellegrini, 2002). For example, among the Santal of India:

> Another pantomimic game is sim sim or "chicken, chicken." In this game the boys and girls form a line holding each other, while a boy flaps his arms like a kite. The kite has to pick off the endmost player without itself being caught. As it dives first one way and then another, the line sways violently about and tries to check it. Finally when only one "chicken" is left, there is a great struggle between the two and they throw each other down and wrestle on the ground. (Archer, 1974, p. 36)

VARIATIONS ON THE THEME

We have been considering how a R&T theme recurs across diverse cultures. R&T also is subject to various influences from the social and physical environment. At a micro level, school policy toward R&T can have an effect on how often children engage in it (Aldis, 1975, p. 285). Other micro-environmental influences appear to be linked to R&T. For example, wrestling for superior position is more likely to occur in grassy areas of a playground than on pavement, in larger classrooms than in smaller ones, and in larger groups of children than in smaller groups (Aldis, 1975; Humphreys & Smith, 1984; Pellegrini, 1987, 1988).

At a macro level, cultural features may influence the form and amount of R&T. As reflected in the eHRAF material considered above, children are great imitators of adult behavior. Hoebel (1960, p. 92) writes of the Cheyenne of North America: "A child does not have to wait until he is grown up to be able to practice what is preached and to experience the satisfaction of performance. Cheyenne children are little replicas of their elders in interests and deeds. Children begin to learn adult activities and practice them in play at incredibly early ages."

During enculturation, children acquire representations of what the world is like, adopt particular sets of values, and gain knowledge about the cultural meaning of actions and events. The Semai of Malaysia, for example, are one of the most peaceable cultures in the world (Dentan, 1978; Robarchek, 1980, 1997). Robarchek explains that

> among Semai, the expression of the cultural values of nurturance, affiliation, and non-aggression in the behavior of adults means that children have little opportunity to learn *how* to be aggressive. Semai children have few models of aggressive behavior; they never see adults fighting, never see fathers attacking mothers. Moreover, they themselves are almost never the objects of aggression either by adults or by other children. (1980, p. 113)

Dentan explains that otherwise permissive Semai adults are not accepting of children's aggression:

> . . . the nonviolent image is stated not merely as an ideal but also as a fact. The Semai expect that their children will conform to the image. Children become aware of this expectation as it is manifested both in subtle daily ways and in the open shock of adults when a child loses its temper. In the latter case, an adult immediately snatches up the angry child and carries it off wailing to its house. This abrupt intervention is probably all the more frightening because adults usually are indifferent to children's activities. (1968, p. 61)

The shunning of physical aggression by the Semai seems to spill over onto R&T. Robarchek (personal communication, December 1998) reports that during fieldwork among the Semai, he saw "very little 'playfighting'." Dentan (1978, p. 132; 1968, p. 59; personal communication, September 1998) describes two play-fighting games among the nonviolent Semai that are notable for their mildness compared to typical R&T in other more aggression-tolerant cultures such as the United States. In the first, "two children, often of disparate sizes, put their hands on each other's shoulders and wrestle, giggling, but never quite knocking each other over" (Dentan, 1978, p. 132). The other kind of Semai play fighting consists of pairs of children in the 3- to 12-year age range flailing at each other with sticks, but stopping just before hitting each other, an activity that "generates a lot of excitement" (Dentan, 1978, p. 132). Dentan (1968, p. 59) comments that "most Semai children's games are small scale rehearsals of adult activity;" and more specifically, the stick flailing game "seems to be a sort of symbolic rehearsal for refraining from violence." Dentan (personal communication, September 1998) reports seeing a great deal of chasing and fleeing, which is a mild, noncontact form of R&T (see Pellegrini, 1994). It would seem that Semai children incorporate the highly valued nonviolent patterns of their culture into their R&T, resulting in particularly tame forms of play fighting in comparison to most other societies.

A comparison of two Zapotec communities provides quantitative support for the proposition that R&T is influenced by cultural values learned during socialization. The communities, referred to by the pseudonyms San Andrés and La Paz, are situated about 6 kilometers apart in the semiarid Valley of Oaxaca in southern Mexico (Fry, 1988, 1992, 2004, in press). San Andrés and La Paz are similar in many respects—both places are populated by indigenous peasant farmers subsisting largely on maize, beans, and squash—but there also are differences. The people of San Andrés are more aggressive than their neighbors in La Paz (Fry, 1992, 1994, 2004). Parents in San Andrés physically punish their children more often and more severely than do La Paz parents; San Andrés parents advocate the use of physical punishment significantly more often than do La Paz parents, who instead favor positive verbal responses to child misconduct (Fry, 1993). The homicide rate in San Andrés is about five times greater than the low rate for La Paz (equivalent to 18.1 versus 3.4 per 100,000 persons per year; Paddock, 1982, and personal communication, February 1986). Fistfights, usually involving inebriated individuals, can be observed in San Andrés with some regularity, whereas comparable fights in La Paz are rare.

R&T is viewed more negatively in La Paz than in San Andrés and is discouraged by adults to some degree. On several occasions La Paz parents were observed telling their children to stop play fighting. O'Nell (1969, p. 251), who also worked in La Paz, reports that of 21 of the La Paz fathers he questioned, only four said that they approved of play fighting; the majority disapproved of both play fighting and fighting.

The divergent patterns of behavior among teenage boys also reflect different attitudes about R&T in the two places. Horseplay is a regular occurrence among San Andrés teens, whereas it occurs very infrequently among La Paz teens. The difference in frequency of play fighting between teenagers in San Andrés and La Paz is magnified further for adults. Women tend not to engage in horseplay in either place, but San Andrés men punch and slap each other as a from of greeting or teasing, engage in mock fights, steal each other's hats, swear at each other in jest, and so forth. Men from La Paz do not act in this manner. They refer to this type of conduct in their San Andrés neighbors as disrespectful. For example, a La Paz man who regularly visited San Andrés called them "unfriendly, egotistical barbarians, who are always swearing" (Fry, 1994, p. 140).

Systematic behavior observations of samples of 3- to 8-year-old focal children revealed that intercommunity differences also existed in children's behavior (Fry, 1988, 1992). San Andrés children engaged in twice as much aggression as La Paz children (0.78 vs. 0.39 episodes/hour), a significant difference. A series of differences between the communities regarding play aggression is presented in Table 4.2. First, the children from San Andrés, compared to those of La Paz, had significantly higher rates and durations of play fighting. Second, the San Andrés children responded to play fighting directed at them with a play response significantly more often than did the La Paz children. This suggests a more common mutual wish among San Andrés children than among La Paz children to participate in play fighting.

TABLE 4.2. Differences in R&T between San Andrés and La Paz

Feature	San Andrés (n = 24)	La Paz (n = 24)
Rate of R&T (episodes/hour)		
Mean	6.9	3.7
Standard deviation	6.8	4.0
Duration of R&T (minutes/hour)		
Mean	2.33	0.88
Standard deviation	3.5	2.1
Rate of response to R&T (rate of response/overall rate)	0.33	0.24
Rates of R&T (episodes/hour) in younger and older children		
3–5 years	3.7	2.7
6–8 years	10.2	4.6
Increase in rates with age and significance of increase	6.5 (10.2–3.7 = 6.5) p = .016	1.9 (4.6–2.7 = 1.9) Not significant (p = .27)

Note. All rates and durations pertain to samples of focal children in the two communities, as described in Fry (1988, 1992).

Third, when each sample was split into younger (3–5 years) and older (6–8 years) subgroups, the older San Andrés children were found to engage in significantly more play aggression than the younger San Andrés group. A similar trend was apparent in La Paz; however, the difference between the age groups was relatively small and not statistically significant. In other words, over the 3–8-year age range, the rate of play aggression significantly increased in San Andrés but not in La Paz.

The sex and the community membership of play initiators and play recipients also reflects an interesting pattern. Table 4.3 presents sex-dyad comparisons for children in each community independently. San Andrés boys both initiate and receive more play fighting than any other subgroup, followed by San Andrés girls, La Paz girls, and finally, *La Paz boys.* The La Paz pattern, although not a large difference, is opposite of that in San Andrés and in most other studies (e.g., Humphreys & Smith, 1987; Pellegrini, 1987, 2002). Girls in both communities were significantly more likely to participate in play fighting with other girls than with boys. How-

TABLE 4.3. Initiators and Recipients of R&T by Sex in San Andrés and La Paz

	No. of episodes (% within each community)		Binomial expansion	
Community	Sex of initiator	Sex of recipient	z	Two-tailed p
San Andrés	Male: 269 (59%)	Male: 211 (46%) Female: 58 (13%)	9.33	< .00001
San Andrés	Female: 185 (41%)	Male: 53 (12%) Female: 132 (29%)	5.81	< .00001
Subtotal	454 (100%)			
La Paz	Male: 104 (38%)	Male: 55 (20%) Female: 49 (18%)	0.58	0.56
La Paz	Female: 168 (62%)	Male: 69 (25%) Female: 99 (36%)	2.31	0.02
Subtotal	272 (100%)			

Note. The episode numbers reflect weighting of the data to correct for slight differences in observation time between girls and boys and between the two communities (see Fry, 1990).

ever, whereas boy-to-boy play fighting was significantly more common than boy-to-girl play fighting in San Andrés, a corresponding preference for boy-to-boy partners did not exist in La Paz. Boys in La Paz initiated play aggression in nearly equal frequencies to girls and boys.

Against the backdrop of the peaceful ethos and behavior of La Paz, these findings suggest an intriguing possibility. The people of La Paz are aware that, generally speaking, male aggression is potentially more dangerous than female aggression. One La Paz woman, when asked about sex differences, said "Men are the ones that kill" (although not often in La Paz). Does the lower level of play fighting among La Paz boys than among La Paz girls reflect stronger socialization pressures aimed against play fighting in boys in this peaceful community? Do the socializing adults in La Paz, perhaps recognizing the greater dangers in male aggression, more actively discourage play fighting, along with fighting, in boys than in girls? Recall O'Nell's (1969) finding that a majority of La Paz fathers disapprove of fighting, even in play. Focusing anti-play-fighting socialization on boys, in particular, might account for this result.

The La Paz Zapotec consistently exhibit less aggressive attitudes, expectations, and practices than do their neighbors in San Andrés. Play fighting also is less acceptable in La Paz than in San Andrés, a value that is reflected in statistically significant differences related to play-fighting behavior. By comparison to their San Andrés counterparts, La Paz children engaged in less play fighting in terms of both rates and durations of episodes, were more likely to ignore a playful act directed at them, and did not show a comparable increase in rate of play fighting with age. These comparisons, in combination, show an overall reluctance of La Paz children to participate in play fighting relative to San Andrés children.

These differences can be explained with a socialization model. The children of San Andrés and La Paz imitate the behavior of their elders and increasingly engage in activities that are accepted, expected, and rewarded in their communities. The children of San Andrés witness adult brawls from time to time, observe teenagers fighting and both teens and grown men engaging in horseplay, and see and experience periodic corporal punishment. In La Paz, however, teenagers and adults hardly ever use physical punishment and rarely exchange blows, even in play. In short, the children of La Paz are not normally exposed to the types of aggression and mock aggression that their San Andrés peers periodically witness. It seems quite plausible that Bandura's (1983, p. 5) view of aggression could be extended to R&T as well: "The specific forms that [play] aggressive behavior takes, the frequency with which it is expressed, the situations in which it is displayed, and the specific targets selected for attack are largely determined by social learning factors." This conclusion is more generally bolstered by observations of many anthropologists (e.g., Baker, 1998, p. 103; Dentan, 1968, p. 59; Gorer, 1967; Hoebel, 1960, p. 92; Koch, 1974, Plates 12 & 13, pp. 52–53; Tonkinson, 1978, p. 66; and the eHRAF findings discussed above) that the play of children replicates the activities of their elders.

INTERPRETATION OF ADAPTIVE FUNCTION

The apparent universal distribution of R&T across cultures suggests that it evolved to serve an adaptive function or functions. The consideration of evolutionary function has caught the attention of many R&T researchers (Boulton, 1991a, 1991b, 1994, 1996; Fry, 1987, 1990; Humphreys & Smith, 1987; Neill, 1976, 1985; Pellegrini, 1994, 2002, 2003; Smith, 1982, 1997; Smith & Boulton, 1990). Three of the most common functional hypotheses are that R&T may (1) facilitate the development of social skills, (2) allow the practice under safe conditions of fighting skills, and (3) contribute to the establishment and maintenance of immediate or long-term dominance (Pellegrini, 2002; Smith, 1997). The social skills hypothesis garners the least support of the three because it provides, for instance, no explanation for sex differences in R&T (Pellegrini, 2002). By contrast, the existence of sex differences in R&T that parallel sex differences in adult aggression and physical competition are congruent with both the fighting practice and the dominance hypotheses.

The fighting practice and dominance hypotheses may be applicable to different degrees at different stages of development. Even at the same age, these two functional explanations need not be viewed as mutually exclusive (see Neill, 1985). For example, assessing a partner's strength during play fighting could relate to gaining practice in a generalizable skill useful in later life (Boulton, 1996; Fry, 1990) and to establishing dominance over a particular opponent in the short or long term (Pellegrini, 2002).

In young children, R&T is rarely exploited for aggressive ends. With age, the abuse of the playful context to exert dominance seems to become more likely (Pellegrini, 2003; Smith et al., 1992). Several events observed among Zapotec teenagers indicate that the clear distinction between play fighting and fighting, apparent in the 3- to 8-year-old Zapotec children, may blur somewhat by the teenage years. Sometimes horseplay among teenagers appeared, judging from facial expressions, to become somewhat serious, but then shifted back to play again (Fry, 1987). This observation supports Neill's (1976, pp. 218, 219) inferences that the intermingling of

R&T and aggression sometimes occurs among 12- to 13-year-old boys: "Once the weaker boy has registered distress the bond can be maintained by the fight taking a more playful form, but if he does not do so at the start of the fight, the stronger boy may increase the intensity of the fight until he does." In sum, by adolescence, R&T may have a dominance function, an interpretation that receives support from observational and interview studies with older children (Pellegrini, 1994, 2002, 2003; Smith et al., 1992).

When a playing child harms a partner, the harm may result from an honest mistake—an accident. Alternatively, an individual might attempt to cheat—that is, to exploit the situation by exerting dominance or causing an injury under the guise of play (Fagen, 1981). However, a third possibility may be applicable to the rough, dominance-related form of R&T displayed by adolescent boys. By this age, R&T participants seem to be well aware that play and aggression are not as distinct phenomena as they were at younger ages (e.g., Smith et al., 1992, 2002). To play with Bateson's (1972) idea of metacommunication, instead of the shared understanding between older partners being "this is not real aggression," as would seem to aptly apply to the R&T of young children, the shared understanding becomes something like "we both know this has a serious component, but we implicitly agree to pretend that it does not," or, perhaps, "we both know that our status is involved, but it is not as serious a situation as it would be if we were to stop pretending that it is not serious." Perhaps some additional insights about the likely functions of R&T might arise from a consideration of adult aggression and competitive contests (Fry, 1990), especially within the forager type of social organization that has characterized human existence over long expanses of evolutionary time.

PATTERNS OF AGGRESSION AND RESTRAINT

Modern humans possess a legacy of adaptations acquired over many millennia. Until about 10,000 years ago, humans and their ancestors lived as nomadic hunter-gatherers. In evolutionary terms, the availability of projectile weaponry is probably very recent. The bow and arrow may have arisen some 45,000 years ago, but perhaps not until as recently as 17,000 to 10,500 years ago—in either case, not long ago in relation to the two million years that the genus *Homo* has existed (Smith, 1999). Presumably, wooden spears have a much longer history, but being of perishable material, have not been preserved in the archeological record. On a daily basis, weapons were used for hunting, but occasionally they were directed at persons (Roper, 1969).

As a supplemental approach to archaeology, an assessment of recurring patterns across current-day nomadic hunter-gatherer societies can provide a basis for drawing inferences about the human past (Boehm, 1999; Fry, in press). Based on observations of extant nomadic foragers, the rates of aggression in the past probably were considerably variable from one group to the next. Much of the aggression that did occur probably did not involve the use of weapons, but less frequently, attacks probably did involve rocks, stone blades, clubs, or spears. Third-party peacemakers and conflict-resolution mechanisms are regular features of extant hunter-gatherer social life and no doubt also existed in the past (Boehm, 1999; Fry, 2000; Gardner, 2004; Lee, 1993; Tonkinson, 1978, 2004).

Nomadic hunter-gatherers live in bands whose composition varies as people transfer among groups. Nomadic foragers have few material possessions, are politically egalitarian, and tend to be widely dispersed. Disputes tend to be personal, not collective (Reyna, 1994; Service, 1966). Nomadic hunter-gatherer bands tend not to engage in war or have militaristic value orientations (Fry, in press; Kelly, 1995, Table 8.1; Steward, 1968). Moreover, the amount of brawling and homicide within forager band society varies from one to the next. Some groups, such as the Paliyan of India (Gardner, 2004) and the Batek Semang of Malaysia (Endicott, 1979), have nonviolent ideals and physical aggression is extremely rare, whereas other band societies experience regular fighting and periodic killings (e.g., Tonkinson, 1978, 2004). Based on an extensive survey of ethnographic accounts, Boehm (1999) notes an overarching pattern wherein the members of band societies do not tolerate overly aggressive persons. Recidivist killers and otherwise dangerous individuals are likely to be executed for the public good (Balikci, 1970; Boehm, 1999; Lee, 1993). Lacking authoritative leadership, courts, police, and mental hospitals, nomadic band societies nonetheless manage to deal with much conflict through avoidance, discussion, group meetings, contests, ostracism, and other nonviolent or aggression-limiting ways (Boehm 1999; Fry, 2000, in press). Nomadic foragers are famous for voting with their feet. In short, the aggression in nomadic band society is not as rampant as is sometimes assumed.

The somewhat ambiguous intermingling of play and aggression among adolescents has a parallel in the contests that occur in some forager bands. First, like adolescent R&T, contests for sport, "for fun," can include a dominance component. Second, contests as a form of dispute resolution also can involve dominance struggles while allowing for the resolution of differences with less injury than might occur during real aggression. Both types of contests are periodically reported among nomadic foragers and in other kinds of societies, and both types have features that straddle play and aggression.

Contests have rules that promote restraint, and spectators take a role in enforcing the rules, if necessary. Across cultural settings, the metacommunicative context of contests is that they are simultaneously "serious yet not serious," or at least not as serious as unbridled aggression. In this way, contests directly parallel the R&T of adolescents, in that both may vacillate back and forth between play and aggression in the gray area of exercising restraint while simultaneously seeking to dominate an opponent. Winning by the rules enhances esteem, but winning through cheating may have an opposite effect when the spectators and the social group are one and the same.

Based on a survey of Inuit band societies, Hoebel (1967, p. 92) points out how contests serve as a means for resolving disputes without the loss of life:

> Homicidal dispute, though prevalent, is made less frequent in many Eskimo groups by recourse to regulated combat—wresting, buffeting, and butting. . . . The object of the boxing and butting contests is not annihilation, but subjection. . . . Whatever the facts underlying the dispute, they are irrelevant to the outcome. The man who wins, wins social esteem. He who loses, suffers loss of social rank.

Among the Netsilik Inuit of the central Canadian Arctic, interpersonal conflicts were handled in several ways in traditional society (Balikci, 1970). Avoidance was commonly practiced. Overall, fighting was prevented due to a set of concerns: people feared aggressive retaliation, the loss of beneficial relationships, and sorcery by an opponent. Moreover, the Netsilik utilized both physical and verbal ritualized contests to settle disputes without serious danger to the participants. These contests had definite rules. In a stylized physical fight, a man who challenged another had to receive the first blow (Balikci, 1970). The two opponents stripped to the waist and stood opposite each other. Back and forth, in turn, each struck the other one time, using blows directed at either the forehead or a shoulder. This exchange continued until one contestant gave up. A Netsilik man explained that "after the fight, it is all over; it was as if they had never fought before" (Balikci, 1970, p. 186).

Verbal song duels also were used to settle disputes. Under the rules of song dueling, opponents had a free range to blast their antagonists in verse. In some cases, both parties were satisfied to have been able to reciprocally insult and deride their opponents publicly, so a dispute was brought to a close. If anger remained, the disputants might move on to a fighting contest. Balikci (1970) reports that the outcome of the physical contest settled the matter once and for all.

Moving to the opposite end of the planet, the nomadic hunting and gathering Ona of Tierra del Fuego engaged in wrestling contests both for dispute resolution and as a sport before their traditional customs were disrupted by outside influences. The Ona material shows that boys practiced wrestling in imitation of their elders (also see description quoted above): "Boys of all ages already amuse themselves with it, the candidates of the Kloketen celebration must practice it, and ambition drives younger men to distinguish themselves in this as much as possible" (Gusinde, 1931, p. 1624). Thus wrestling clearly involved a dominance element. And, as among the Netsilik, wrestling provided a relatively safe context in which to settle disputes or to assert dominance:

> It is carried on, to be sure, in the same way as ordinary wrestling for sheer fun, but here it proceeds with increased exertion and more malice. The occasions are insult, defamation of honor, or slighting another man, who will not put up with such things. . . . [The wrestling] happens only if each believes he is a match for his opponent; otherwise the weaker one avoids challenging the other to fight.
>
> . . . The two move toward each other and seize each other tightly. The previous irritation and the heightened jealousy cause each to attack boldly; they summon their utmost strength and plant themselves against each other in desperate rage, until finally one must succumb, either by being pressed against a tree or thrown on the ground. With this the existing disagreement has been settled to some extent, at least for today, namely, to the disadvantage of the one defeated. . . . One who had to leave as the one defeated took this dishonor very seriously; his people also often reminded him of it. (Gusinde, 1931, pp. 645–646)

Wrestling for sport was similar, but less intense, and engaged in by boys and men alike. The competitive element was clearly visible among boys: "The example of the adults often spurs them on, for, when an opportunity presents itself, they also valiantly fall to and never tire in their rivalry to down an opponent" (Gusinde, 1931, p. 1623). As among the Netsilik, the Ona wrestling contests had rules:

> To trip up a person would be impermissible. They choose soft earth, preferably dry mossy ground, where women and children and watching men form a large circle, in the inner space of which the wrestlers step. The latter always throw off their cloaks, so that the play of their tensed muscles on their naked bodies and the suppleness of all their movements can be observed with great pleasure by all present.

> . . . The wrestling of two men in order to test their own strength, is most frequently practiced. For the participants, this is an exercise and sporting performance; for the spectators, entertainment rich in enjoyment. . . . Although the two opponents seize each other resolutely and gradually increase their efforts to the utmost, the wrestling never degenerates into ill-feeling, even though the stronger one finally knocks the weaker one to the ground with such force that he can sometimes get up only with difficulty. Many a man may be badly hurt, to be sure, but in the long run he may not escape further fights, but must venture on to increasingly stronger opponents in order not to get the reputation of a weak coward. For a very long time afterwards such a performance stimulates constant discussions, comparisons, and various opinions. (Gusinde, 1931, pp. 1624–1625)

Clearly, a couple of ethnographic examples from nomadic hunter-gatherer societies cannot prove particular functional interpretations of R&T to be correct, but perhaps they offer additional relevant clues for assessing the plausibility of the practice and dominance hypotheses. First, the parallels in design features between the R&T of children and the adult contests are inescapable. Furthermore, R&T may also have some parallels in the real aggression of adults, as reflected in the following description of a life-and-death struggle among the Netsilik:

> As he knelt down at the water, Ikpagittoq attacked him from behind, trying to throw him to the ground. A struggle developed, while Oksoangutaq stood by watching until the embattled Ikpagittoq shouted at him, "You said you wanted to kill this man, what are you waiting for?" Oksoangutaq stepped up and pushed his knife into Saojori's neck, killing him on the spot. (Balikci, 1970, pp. 180–181)

The R&T pattern described for human children across diverse cultural settings contains obvious similarities to both restrained fighting (as in contests) and unrestrained aggression (as in the above quotation). The design is similar (see Fry, 1990). The behavioral elements, gross motor patterns, and the goal of immobilizing an opponent (temporarily or permanently) apparent during contests and more serious fighting have parallels in the R&T of human children. These correspondences are congruent with the hypothesis that human play fighting serves a practice function (Fry, 1990; Neill, 1985; Pellegrini, 2002; Smith, 1982; Smith et al., 1992).

At the same time, the adult contests in both the Netsilik and Ona cases clearly involve dominance struggles. The literature on R&T suggests that an increasing dominance dimension to play fighting comes into play with increasing age, at least in boys (Pellegrini, 2002, 2003). In adolescents, dominance may have immediate benefits within the peer group as well as potential longer-term benefits (Neill, 1985). The clear link in the foregoing examples between winning a contest and increasing one's status (or losing and suffering a loss in status) hints at the importance of dominance striving in male humans. In adult contests, as in R&T among adolescents, curtailed aggression parallels the ritualized aggression of some animal species. Restrained aggression allows dominance to be established with substantially less risk to the participants than would result from all-out fighting. The metacommunicative parallels between adolescent R&T and adult contests as "serious but not serious" are intriguing. Could adolescent R&T provide practice at participating in restrained, rule-based competitive struggles later in life? If so, we may have a convergence, in adolescence, of practice and dominance functions of R&T. The ways in which contests and adolescent R&T share with each other intermediate features between play and aggression are summarized in Table 4.4.

TABLE 4.4. A Comparison of R&T, Contests, and Adult Aggression

Preadolescent R&T	Adolescent R&T and contests	Adult aggression
Winning is *not* important	Winning is important (there may be sporting or juridical elements to contests)	Winning is important
Friendly	Friendly-to-hostile	Hostile
No attempt to dominate	May include domination attempts	Attempts to dominate
No intent to injure	No intent to seriously injure	Intent to injure
Serious injury extremely rare	Serious injury possible but rare	Serious injury possible
Much restraint	Some restraint	Minimal restraint
Resources irrelevant	Resources possibly relevant	Resources usually relevant
Others not interested	Others interested as referees and spectators	Others interested as peacekeepers or allies

Note. The patterns suggested in this table are derived from the literature on R&T, as reviewed in this chapter, and from surveying numerous ethnographic accounts of conflict management and aggression among nomadic hunter-gatherers and in other societies.

CONCLUSIONS

Certain common elements of human R&T are apparent in widely distributed cultures. When information exists, wrestling and grappling, restrained hitting and punching, and chasing and fleeing recur in descriptions of children's R&T. These behaviors usually occur in conjunction with laughs, smiles, play faces, and sometimes with playful vocalizations or speech indicating the fantasy elements of play. The cross-cultural data are limited both in number of relevant sources and in the detail of reporting, but the available evidence reflects these common features. An additional aspect of the pattern is that boys tend to engage in more R&T—especially in the roughest forms, such as wrestling—than do girls. Although data are not available for many cultures, the evidence suggests that a tendency among children, especially boys, to engage in R&T is probably a human universal. At the same time, cultural variations on the R&T theme exist. Fantasy elements reflect cultural features, such as when British children shoot imaginary guns, when Zapotec children play at rodeo by attempting to subdue a "bull" (played by another child), or when Santal children play the *sim sim* game of "chickens" versus a "kite." In some social circumstances, as among the Semai and La Paz Zapotec, adults discourage children from engaging in R&T. The milder forms of R&T among the Semai and the relatively fewer incidences of R&T among La Paz Zapotec compared to San Andrés Zapotec illustrate that R&T is open to various social influences.

Research on R&T in children suggests that both practice and dominance functions are likely. For humans, the practice of fighting skills may be a more complex task than in some other species. Humans may have to practice not only fighting maneuvers but also restraint within a social world that includes rules for fighting and dispute resolution. It is interesting to contemplate how much aggression among humans is restrained, curtailed, or limited through social conventions, enforced by the participants themselves and by other members of the group. Contests are one obvious example of restrained aggression and may be similar to the R&T in adolescent boys by providing a way to resolve dominance struggles with minimal risk to the participants. At the same time, contests are by no means a universal solution to this problem; humans generally, and nomadic foragers in particular, use many ways to handle conflict, some more competitive than others (Fry, 2000). As suggested by the data on the La Paz Zapotec and the Semai, and reinforced by the existence of highly peaceful nomadic hunter-gatherer societies such as the Batek and the Paliyan, the expression of both aggression and R&T can be reduced to very low levels. At the same time, the manifestation of the R&T theme in children around the world suggests the presence of evolved functions for the behavioral pattern, of which the practice of fighting skills, broadly conceived, and dominance assertion seem likely candidates.

ACKNOWLEDGMENTS

I would like to thank editors Tony Pellegrini and Peter K. Smith for their helpful comments on an earlier draft of this chapter. John Paddock kindly provided unpublished homicide data for Zapotec communities in Mexico. Bob Dentan and Clay Robarchek graciously responded to queries about the nature of rough-and-tumble play among the Semai of Malaysia, providing information for which I am most grateful. I thank the people of San Andrés and La Paz, especially the focal children and their parents, for their helpful cooperation and hospitality. Some of the findings reported in this chapter were gathered as part of projects funded by the National Science Foundation (Grant Nos. 8117478 and 0313670), the Wenner–Gren Foundation for Anthropological Research (Grant No. 4117), and the United States Institute of Peace (Grant No. 02399F), whose support is gratefully acknowledged.

REFERENCES

Aldis, O. (1975). *Play fighting.* New York: Academic Press.

Archer, G. W. (1974). *The hill of flutes—life, love, and poetry in tribal India: A portrait of the Santals* (eHRAF version, Santal, Doc. 6, 1998). Pittsburgh, PA: University of Pittsburgh Press.

Baker, V. J. (1998). *A Sinhalese village in Sri Lanka.* Fort Worth, TX: Harcourt Brace.

Balikci, A. (1970). *The Netsilik Eskimo.* Garden City, NY: Natural History Press.

Bandura, A. (1983). Psychological mechanisms of aggression. In R. Geen & E. Dunnerstein (Eds.), *Aggression: Theoretical and empirical reviews* (pp. 1–35). Englewood Cliffs, NJ: Prentice-Hall.

Bateson, G. (1972). A theory of play and fantasy. In G. Bateson (Ed.), *Steps to an ecology of mind* (pp. 177–193). Northvale, NJ: Aronson.

Bekoff, M. (1981). Development of agonistic behavior: Ethological and ecological aspects. In P. F. Brain & D. Benton (Eds.), *Multidisciplinary approaches to aggression research* (pp. 166–177). Amsterdam: Elsevier.

Bekoff, M., & Byers, J. (1981). A critical reanalysis of the ontogeny and phylogeny of mammalian social and locomotor play: An ethological hornet's nest. In K. Immelmann, G. W. Barlow, L. Petrinivich, & M. Main (Eds.), *The Bielefeld interdisciplinary project* (pp. 296–337). Cambridge, UK: Cambridge University Press.

Bekoff, M., & Byers, J. (1985). The development of behavior from evolutionary and ecological perspectives in mammals and birds. In M. Hecht, W. Bruce, & G. T. Prance (Eds.), *Evolutionary biology* (Vol. 19, pp. 215–286). New York: Plenum Press.

Blurton Jones, N. G. (1967). An ethological study of some aspects of social behaviour of children in nursery school. In D. Morris (Ed.), *Primate ethology* (pp. 347–368). London: Weidenfeld & Nicolson.

Blurton Jones, N. G. (1972). Categories of child–child interaction. In N. G. Blurton Jones (Ed.), *Ethological studies of child behaviour* (pp. 97–127). Cambridge, UK: Cambridge University Press.

Blurton Jones, N. G., & Konner, M. J. (1973). Sex differences in behaviour of London and Bushman children. In R. Michael & J. Crook (Eds.), *Comparative ecology and behaviour of primates* (pp. 690–750). New York: Academic Press.

Boehm, C. (1999). *Hierarchy in the forest: The evolution of egalitarian behavior.* Cambridge, MA: Harvard University Press.

Boulton, M. J. (1991a). A comparison of structural and functional features of middle school children's playful and aggressive fighting. *Ethology and Sociobiology, 12,* 119–145.

Boulton, M. J. (1991b). Partner preferences in middle school children's playful fighting and chasing: A test of some competing functional hypotheses. *Ethology and Sociobiology, 12,* 177–193.

Boulton, M. J. (1993a). Children's abilities to distinguish between playful and aggressive fighting: A developmental perspective. *British Journal of Developmental Psychology, 11,* 249–263.

Boulton, M. J. (1993b). Proximate causes of aggressive fighting in middle school children. *British Journal of Educational Psychology, 63,* 231–244.

Boulton, M. J. (1994). The relationship between playful and aggressive fighting in children, adolescents and adults. In J. Archer (Ed.), *Male violence* (pp. 23–41). London: Routledge.

Boulton, M. J. (1996). A comparison of 8- and 11-year-old girls' and boys' participation in specific types of rough-and-tumble play and aggressive fighting: Implications for functional hypotheses. *Aggressive Behavior, 22,* 271–287.

Boulton, M. J., & Smith, P. K. (1992). The social nature of playfighting and playchasing: Mechanisms and strategies underlying co-operation and compromise. In J. H. Barkow, L. Cosmides, & J. Tooby (Eds.), *The adapted mind* (pp. 429–444). New York: Oxford University Press.

Camras, L. (1980). Animal threat displays and children's facial expressions: A comparison. In D.R. Omark, F. F. Strayer, & D. G. Freedman (Eds.), *Dominance relations: An ethnological view of human conflict and social interaction* (pp. 121–136). New York: Garland STPM.

Costabile, A., Smith, P. K., Matheson, L., Aston, J., Hunter, T., & Boulton, M. J. (1991). A cross-national comparison of how children distinguish serious and playful fighting. *Developmental Psychology, 27,* 881–887.

Dentan, R. K. (1968). *The Semai: A nonviolent people of Malaya.* New York: Holt, Rinehart, & Winston.

Dentan, R. K. (1978). Notes on childhood in a nonviolent context: The Semai case (Malaysia). In A. Montagu (Ed.), *Learning non-aggression: The experience of nonliterate societies* (pp. 94–143). Oxford, UK: Oxford University Press.

Eibl-Eibesfeldt, I. (1974). The myth of the aggression-free hunter and gatherer society. In R. L. Holloway (Ed.), *Primate aggression, territoriality, and xenophobia: A comparative perspective* (pp. 425–457). New York: Academic Press.

Einon, D., & Potegal, M. (1991). Enhanced defense in adult rats deprived of playfighting experience as juveniles. *Aggressive Behavior, 17*, 27–40.

Ekman, P., Friesen, W. V., O'Sullivan, M., Chan, A., Diacoyannitarlatzis, I., Heider, K., Krause, R., LeCompte, W. A., Pitcairn, T., Riccibitti, P. E., Scherer, K., Tomita, M., & Tzavaras, A. (1987). Universals and cultural-differences in the judgments of facial expressions of emotion. *Journal of Personality and Social Psychology, 53*, 712–717.

Endicott, K. (1979). *Batek Negrito religion: The world-view and rituals of a hunting and gathering people of Peninsular Malaysia.* Oxford, UK: Clarendon Press.

Fagen, R. M. (1978). Evolutionary biological models of animal play behavior. In G. Burdghardt & M. Bekoff (Eds.), *The development of behavior: Comparative and evolutionary aspects* (pp. 385–404). New York: Garland.

Fagen, R. M. (1981). *Animal play behavior.* New York: Oxford University Press.

Frey, C., & Hoppe-Graff, S. (1994). Serious and playful aggression in Brazilian girls and boys. *Sex Roles, 30*, 249–268.

Fry, D. P. (1987). Differences between playfighting and serious fighting among Zapotec children. *Ethology and Sociobiology, 8*, 285–306.

Fry, D. P. (1988). Intercommunity differences in aggression among Zapotec children. *Child Development, 59*, 1008–1019.

Fry, D. P. (1990). Play aggression among Zapotec children: Implications for the practice hypothesis. *Aggressive Behavior, 16*, 321–340.

Fry, D. P. (1992). "Respect for the rights of others is peace": Learning aggression versus non-aggression among the Zapotec. *American Anthropologist, 94*, 621–639.

Fry, D. P. (1993). Intergenerational transmission of disciplinary practices and approaches to conflict. *Human Organization, 52*, 176–185.

Fry, D. P. (1994). Maintaining social tranquility: Internal and external loci of aggression control. In L. E. Sponsel & T. Gregor (Eds.), *The anthropology of peace and nonviolence* (pp. 133–154). Boulder, CO: Reinner.

Fry, D. P. (2000). Conflict management in cross-cultural perspective. In F. Aureli & F. B. M. de Waal (Eds.), *Natural conflict resolution* (pp. 334–351). Berkeley: University of California Press.

Fry, D. P. (2004). Multiple paths to peace: The "La Paz" Zapotec of Mexico. In G. Kemp & D. P. Fry (Eds.), *Keeping the peace: Conflict resolution and peaceful societies around the world* (pp. 73–87). New York: Routledge.

Fry, D. P. (in press). *The human potential for peace: Challenging the war assumption.* New York: Oxford University Press.

Gardner, P. M. (2004). Respect for all: The Paliyans of South India. In G. Kemp & D. P. Fry (Eds.), *Keeping the peace: Conflict resolution and peaceful societies around the world* (pp. 53–71). New York: Routledge.

Gorer, G. (1967). *Himalayan village: An account of the Lepchas of Sikkim* (2nd ed.). New York: Basic Books.

Grinnell, G. B. (1961). *Pawnee, Blackfoot, and Cheyenne: History and folklore of the Plains.* New York: Scribner's.

Gusinde, M. (1931). *The Fireland Indians* (Vol. 1). Vienna: Verlag der Internationalen Zeitschrift.

Harlow, H. F., & Harlow, M. K. (1965). The affectional systems. *Behavior of Non-Human Primates, 2*, 287–334.

Hoebel, E. A. (1960). *The Cheyennes: Indians of the Great Plains.* New York: Holt, Rinehart & Winston.

Hoebel, E. A. (1967). *The law of primitive man: A study in comparative legal dynamics.* Cambridge, MA: Harvard University Press.

Humphreys, A. P., & Smith, P. K. (1984). Rough-and-tumble in preschool and playground. In P. K. Smith (Ed.), *Play in animals and humans* (pp. 241–266). Oxford, UK: Blackwell.

Humphreys, A. P., & Smith, P. K. (1987). Rough-and-tumble, friendship, and dominance in schoolchildren: Evidence for continuity and change with age. *Child Development, 58*, 201–212.

Jolly, A. (1985). *The evolution of primate behavior* (2nd ed.). New York: Macmillan.

Kelly, R. L. (1995). *The foraging spectrum: Diversity in hunter-gatherer lifeways.* Washington, DC: Smithsonian Institution Press.

Koch, K.-F. (1974). *War and peace in Jalémó: The management of conflict in highland New Guinea.* Cambridge, MA: Harvard University Press.

Konner, M. J. (1972). Aspects of the developmental ethology of a foraging people. In N. G. Blurton Jones (Ed.), *Ethological studies of child behaviour* (pp. 285–304). Cambridge, UK: Cambridge University Press.

Lee, R. B. (1993). *The Dobe Ju/'hoansi.* Fort Worth, TX: Harcourt Brace.

McGrew, W. C. (1972). *An ethological study of children's behavior.* New York: Academic Press.

Nabuzoka, D., & Smith, P. K. (1999). Distinguishing serious and playful fighting by children with learning disabilities and nondisabled children. *Journal of Child Psychology and Psychiatry, 40*, 883–890.

Neill, S. R. St. J. (1976). Aggressive and non-aggressive fighting in twelve-to-thirteen year old preadolescent boys. *Journal of Child Psychology and Psychiatry, 17*, 213–220.

Neill, S. R. St. J. (1985). Rough-and tumble and aggression in schoolchildren: Serious play? *Animal Behaviour, 33*, 1380–1382.

O'Nell, C. W. (1969). *Human development in a Zapotec community with an emphasis on aggression control and its study in dreams.* Unpublished doctoral dissertation, Department of Anthropology, University of Chicago.

Paddock, J. (1982, October). *Anti-violence in Oaxaca, Mexico: Archive research.* Paper presented at the meeting of the American Society for Ethnohistory, Nashville, TN.

Pearsall, M. (1950). *Klamath childhood and education.* Berkeley: University of California Press.

Pellegrini, A. D. (1987). Rough-and-tumble play: Developmental and educational significance. *Educational Psychologist, 22*, 23–43.

Pellegrini, A. D. (1988). Elementary-school children's rough-and-tumble play and social competence. *Developmental Psychology, 24*, 802–806.

Pellegrini, A. D. (1989). What is a category? The case of rough-and-tumble play. *Ethology and Sociobiology, 10*, 331–342.

Pellegrini, A. D. (1993). Boys' rough-and-tumble play, social competence and group composition. *British Journal of Developmental Psychology, 11*, 237–248.

Pellegrini, A. D. (1994). The rough play of adolescent boys of differing sociometric status. *International Journal of Behavioral Development, 17*, 525–540.

Pellegrini, A. D. (1995). Boys' rough-and-tumble play and social competence: Contemporaneous and longitudinal relations. In A. D. Pellegrini (Ed.), *The future of play theory: A multidisciplinary inquiry into the contributions of Brian Sutton-Smith* (pp. 107–126). Albany: State University of New York Press.

Pellegrini, A. D. (2002). Rough-and-tumble play from childhood through adolescence: Development and possible functions. In P. K. Smith & C. H. Hart (Eds.), *Handbook of childhood social development* (pp. 438–453). Oxford, UK: Blackwell.

Pellegrini, A. D. (2003). Perceptions and possible functions of play and real fighting in early adolescence. *Child Development, 74*, 1552–1533.

Pellis, S. M. (1984). Two aspects of play-fighting in a captive group of Oriental small-clawed otters *Amblonyx cinerea. Zeitschrift für Tierpsychologie, 65*, 77–83.

Pellis, S. M., Field, E. F., Smith, L. K., & Pellis, V. C. (1996). Multiple differences in the play fighting of male and female rats: Implications for the causes and functions of play. *Neuroscience and Biobehavioral Reviews, 21*, 105–120.

Reyna, S. P. (1994). A mode of domination approach to organized violence. In S. P. Reyna & R.E. Downs (Eds.), *Studying war: Anthropological perspectives* (pp. 29–65). Leiden, The Netherlands: Gordon & Breach.

Robarchek, C. A. (1980). The image of nonviolence: World view of the Semai Senoi. *Federated Museums Journal (Malaysia), 25*, 103–117.

Robarchek, C. A. (1997). A community of interests: Semai conflict resolution. In D. P. Fry & K. Björkqvist (Eds.), *Cultural variation in conflict resolution: Alternatives to violence* (pp. 51–58). Mahwah, NJ: Erlbaum.

Roper, M. K. (1969). A survey of the evidence for intrahuman killing in the Pleistocene. *Current Anthropology, 10*, 427–459.

Roscoe, J. (1911). *The Baganda.* London: Macmillan.

Roscoe, J. (1923). *The Bakitara or Bunyoro.* Cambridge, UK: Cambridge University Press.

Schåfer, M., & Smith, P. K. (1996). Teachers' perceptions of play fighting and real fighting in primary school. *Educational Research, 38*, 173–181.

Scott, E., & Panksepp, J. (2003). Rough-and-tumble play in human children. *Aggressive Behavior, 29*, 539–551.

Service, E. R. (1966). *The hunters.* Englewood Cliffs, NJ: Prentice-Hall.

Smith, A. B. (1999). Archaeology and evolution of hunters and gatherers. In R. B. Lee & R. Daly (Eds.), *The Cambridge encyclopedia of hunters and gatherers* (pp. 384–390). Cambridge, UK: Cambridge University Press.

Smith, P. K. (1982). Does play matter? Functional and evolutionary aspects of animal and human play. *Behavioral and Brain Sciences, 5*, 139–184.

Smith, P. K. (1997). Play fighting and real fighting: Perspectives on their relationship.

In A. Schmitt, K. Atswanger, K. Grammer, & K. Schafer (Eds.), *New aspects of human ethology* (pp. 47–64). New York: Plenum Press.

Smith, P. K., & Boulton, M. (1990). Rough-and-tumble play, aggression and dominance: Perception and behaviour in children's encounters. *Human Development, 33*, 271–282.

Smith, P. K., Hunter, T., Carvalho, A. M. A., & Costabile, A. (1992). Children's perceptions of playfighting, playchasing and real fighting: A cross-national interview study. *Social Development, 1*, 211–229.

Smith, P. K., & Lewis, K. (1985). Rough-and-tumble play, fighting, and chasing in nursery school children. *Ethology and Sociobiology, 6*, 175–181.

Smith, P. K., Smees, R., & Pellegrini, A. D. (2004). Play fighting and real fighting: Using video playback methodology with young children. *Aggressive Behavior, 30*, 164–173.

Smith, P. K., Smees, R., Pellegrini, A. D., & Menesini, E. (2002). Comparing pupil and teacher perceptions for playful fighting, serious fighting and positive peer interaction. In J. L. Roopnarine (Ed.), *Conceptual, social–cognitive, and contextual issues in the fields of play: Play and culture studies* (Vol. 4, pp. 235–245). Westport, CT: Ablex.

Sorenson, E. R. (1978). Cooperation and freedom among the Fore of New Guinea. In A. Montagu (Ed.), *Learning non-aggression: The experience of non-literate societies* (pp. 12–30). New York: Oxford University Press.

Steward, J. (1968). Causal factors and processes in the evolution of pre-farming societies. In R. B. Lee & I. DeVore (Eds.), *Man the hunter* (pp. 321–334). Chicago: Aldine.

Stross, B. (1970). *Aspects of language acquisition by Tzeltal children.* Ann Arbor, MI: University Microfilms.

Symons, D. (1978). *Play and aggression: A study of rhesus monkeys.* New York: Columbia University Press.

Tonkinson, R. (1978). *The Mardudjara Aborigines: Living the dream in Australia's desert.* New York: Holt, Rinehart & Winston.

Tonkinson, R. (2004). Resolving conflict within the law: The Mardu Aborigines of Australia. In G. Kemp & D. P. Fry (Eds.), *Keeping the peace: Conflict resolution and peaceful societies around the world* (pp. 89–104). New York: Routledge.

Whiting, B. B., & Edwards, C. P. (1988). *Children of different worlds: The foundation of social behavior.* Cambridge, MA: Harvard University Press.

Whiting, B. B., & Whiting, J. (1975). *Children of six cultures: A psycho-cultural analysis.* Cambridge, MA: Harvard University Press.

PART III

OBJECT PLAY

CHAPTER 5

Object Play in Great Apes

Studies in Nature and Captivity

JACKLYN K. RAMSEY AND WILLIAM C. MCGREW

Many, if not all, mammals, as well as some birds and even reptiles, play (Burghardt, 1998; Fagen, 1981). In contrast, only a few species of animals habitually use tools, and of those, primates do so most often (Beck, 1980; McGrew, 1992), with chimpanzees and orangutans being the most frequent tool users (orangutans are seldom seen to use tools in the wild, but often do so in captivity; Lethmate, 1982). Moreover, it is difficult to capture, precisely and comprehensively, the essence of play. Exaggeration, repetition, innovation, or omission of an act, as well as reordered and abbreviated sequences of acts can all contribute to the play behavior of an animal (Fagen, 1981; Loizos, 1967; McGrew, 1977). Power (2000) compared play actions of humans and animals and scrutinized definitions of object play, tool use, and exploration by animals and human juveniles (see also Pellegrini and Gustafson, Chapter 6, this volume). He asserted that no single, simple definition of play is satisfactory and that its elements easily fit into more than one category. Object play may be confused with exploration and tool use because no clear-cut components distinguish one from the other. Furthermore, there are various ways to subcategorize play, such as solitary versus social, or object versus non-object play. In mammals, a special facial expression (play face) or a special vocalization (laughter) accompanies and signals play.

Although the play face and laughter differ across species, they serve the same function, which is to signal nonagonistic intent to others or to express emotion even when alone.

However, when trying to study object play, definitions may be hazy and may not distinguish the overlap across *object exploration, object manipulation,* and *tool use.* Object play is a problematic concept because it is a subcategory of object manipulation, along with the other subcategories of object exploration and tool use (Loizos, 1967; Power, 2000). Object manipulation, which is any treatment of an object, is closely related to object exploration, which is an act whose intent is to familiarize the user with a novel object (Power, 2000). In object play, the point is to incorporate inanimate objects into the animal's play (Hall, 1998). Finally, according to Beck (1980), tool use occurs when an unattached object from the environment is used to alter the form, position, or condition of another object. Because of categorical overlap, the potential for confusion is great.

It is easiest to look for object play in the young, who must learn how to manipulate objects to make use of them functionally. The obvious evolutionary explanation for the presence of object play in juveniles focuses on its usefulness in providing practice or training for certain behavioral patterns needed as adults, such as food processing or tool use (Hall, 1998; Power, 2000). However, what about the object play that occurs in adult animals, which seems to be harder to explain than juvenile play? Given that many explanations for play focus on object play as a prerequisite to learning specific tasks for the future (however, see Pellegrini & Smith, 2002), how can it be functional in adults? Hall (1998) suggests that object play might provide continuing practice that helps adults perfect or maintain basic motor skills, especially when new variation is required. Regardless of an individual's age, object exploration can be useful when new environmental stimuli emerge. In this scenario, object exploration familiarizes the individual with the new stimulus, and play helps it to master its utility.

Another explanation for play is that it may facilitate peer interaction and physical development in immature animals (Mendoza-Granados & Sommer, 1995). In these terms (except for humans), object play is less common in adults because they have already mastered the information needed to make functional use of an object (Fagen, 1993; Hall, 1998). Equally, regular play may promote interaction between experimental knowledge and skills and neuromuscular maturation in order to produce optimal rates of behavioral and cognitive development.

Many a comparative study has run afoul of imprecise or even unstated distinctions among these related phenomena. The published literature on play in great apes is reviewed below, and much of it is anecdotal. We can find no published study of any species in which object play is distinguished from non-play on the basis of concurrent play signals. Furthermore, we can find no previous primate study of the norms of such object play in the wild.

CAPTIVE CHIMPANZEES AND BONOBOS

The genus *Pan* is humankind's nearest living relation, and so takes pride of place in any comparative analysis of the phylogeny of play. Happily, there is a varied (but sketchy) published literature on the subject for chimpanzees, but there is far less known about bonobos.

Yerkes (1943) was the first to report object play in captive chimpanzees, when he described the daily life of a colony of chimpanzees (*Pan troglodytes*) at what was then the Yerkes Laboratories in Florida. He briefly described the favorite play item of the chimpanzees: a car tire suspended by a chain. The chimpanzees swung on it, tied the chain in a knot, twisted it, chewed on it, and cradled it. The single rubber tire gave the chimpanzees endless hours of entertainment.

Viki was a female chimpanzee raised from infancy in a human home (Hayes, 1951). Whether it was play-mothering a doll or running and jumping on a rubber dog while laughing, Viki played with children's toys much as do human children. This description echoes other accounts of home-reared chimpanzee infants: for example, Joni and Nadia Kohts (Ladygina-Kohts, 2002), Gua and the Kelloggs (Kellogg & Kellogg, 1933), and Lucy and the Temerlins (Temerlin, 1975). In a famous anecdote, Hayes recorded Viki pretending to pull behind her a nonexistent toy with a string. If accurately interpreted, this would be virtual object play (see also Gómez & Martín-Andrade, Chapter 7, this volume).

In what appears to be the first ethological study of free play, Loizos (1969) observed a group of chimpanzees, composed of a male and five females, in London Zoo. She compared the agonistic behavior of aggression and submission to those of a more playful nature. She found that although both sets of behavioral patterns contained the same motor components, play had its own signals, especially the play face. Playful approach resulted in the recipient reacting with similar behavioral patterns, often incorporating objects. Social play that once would have been included in agonistic behavior was now recognized to be another category that sometimes led to low-level agonistic behavior, as in teasing (Adang, 1984).

Systematic research on behavioral development is exemplified by Poti and Spinozzi's (1994) study of four male chimpanzee infants in captivity, which sought to assess their developmental stages compared to those of human infants. At the start of the study, the chimpanzees ranged in age from birth to 17 weeks old. They were tested every 4 weeks by introducing various combinations of objects, such as balls, plastic shapes and rattles, pieces of cloth and string, sticks, wheeled toys, a nursing teat, and a xylophone. Between 12 and 17 weeks old, the chimpanzees started to manually explore the objects. In months 8–14, they were not interested in discovering the object's properties, and only one subject actively used an object to move another. Overall, the results showed that chimpanzee and human infants progress through the same developmental stages, although chimpanzees did not advance as far.

An ethological study of a full-sized naturalistic group of chimpanzees was done by Mendoza-Granados and Sommer (1995) in a colony of 25 apes at the Burgers' Zoo in Arnhem, Netherlands, focusing on the play of individuals. They observed the 11 oldest immature chimpanzees (eight females and three males) as they moved about independently during the 4-month study period. Ages of the focal animals ranged from 2.5 years to 9.5 years. All types of play were recorded: social, solitary, and object. Play bouts (n = 1,067) accounted for 19% of the observation time in the focal group. Play bouts without objects outnumbered object play bouts by 2:1. Play frequency was negatively correlated with age, and play occurred more often between agemates than between adult and infant. Males played more than females, and solitary play was seen in females more than in males.

Immature group-living chimpanzees were the subjects of study of Markus and Croft (1995), who observed 12 individuals living in an outdoor enclosure at the Taronga Zoo in Australia. Seven infants (three males, four females) and five juveniles (three males, two females) were observed for social play, object play, and solitary play. Object play was split into six categories according to type of object: Hessian bag, coconut, bamboo, jam fishing (artificial "termite nests" filled with honey or jam), Kikuyo (manipulations of snake grass with skills used later to make nests), and behavioral patterns involving manipulation and mouthing of specific objects without ingestion. The results showed that manipulatory object play with bamboo developed during infancy and into a skilled usage in the juvenile years; however, the authors failed to mention what the skills entailed. Fern, palm fronds, straw, Hessian bags, coconut shells, and bamboo sticks, along with other objects, were used for social and solitary entertainment. Also, individuals of 6 years and older had fewer interactions and showed the least activity in object play.

In focusing on the development of imitation, Bjorklund, Yunger, Bering, and Ragan (2002) chose three juvenile chimpanzees, ages 5, 7, and 9 years, for study. All were reared at home with human and chimpanzee contact from infancy onward, and they spent most of their day in an outdoor enclosure. The authors studied deferred imitation of the use of several different objects and how the chimpanzees reproduced the actions after a human demonstration. Each of the three chimpanzees increased performance from the baseline imitation tasks, with the youngest chimpanzee being the most proficient. The results suggest that with more practice, the more proficient the chimpanzee becomes and possibly the less play in which he or she is motivated to engage.

In the first study of the three most closely related species of hominoids, Vauclair (1984) compared a human infant, a bonobo (*Pan paniscus*) infant, and a common chimpanzee infant on the complexity of object manipulation and intelligence. All three were from 8 to 11 months in age. The human infant was observed at her home, and the others were observed at the nursery of the Yerkes Primate Research Center, in Atlanta, Georgia. Objects used in this study consisted of four nested cubes, two sticks, a plate, doll, and cup. Manipulation of an object was termed active manipulation and was split into five categories: hold, remove, push or pull, wave or shake, and explore. The human infant showed less apprehension about the object and explored it more than did the bonobo or chimpanzee. The human infant also was more proficient in discovering the potential uses of the toy, followed by the bonobo and then the chimpanzee. The chimpanzee and bonobo less often removed the object from the background and more often mouthed it than did the human infant.

Finally, there has been one direct comparison using ethological methods: Takeshita and Walraven (1996) compared captive bonobos and chimpanzees on their object manipulation. Twenty-seven chimpanzees from Burgers' Zoo in Arnhem and eight bonobos from Planckendael Animal Park in Muizen-Mechelen, Belgium, were the subjects of study. They were given four types of objects: wooden spoon, metal bowl, plastic box, and cloth towel. These specific objects were chosen to encourage the manipulation of more than one object at once. The study's main goal was to compare one-handed versus two-handed, symmetrical versus asymmetrical and orienting manipulation between bonobos and chimpanzees. Overall, the nine younger chimpanzees showed many more bouts of object manipulation ($n = 701$) than did the 16 older individuals ($n = 503$). As a species, the bonobos showed a greater tendency for multiple object manipulations than did the chimpanzees. These results suggest, yet again, that younger apes are more curious about novel objects than are adults, and that there may be species differences even between congeneric taxa.

WILD CHIMPANZEES AND BONOBOS

Object play in wild chimpanzees and bonobos seems to consist mainly of manipulating vegetation, such as branches and leaves, but live animals are also of interest.

The first study of object play (*sensu strictu*) in wild chimpanzees was done by McGrew (1977), who followed six wild individuals living at the Gombe National Park in Tanzania. The subjects were three females and three males ranging in age from 1.5 to 5 years. All of Gombe's chimpanzees were able to use probes to fish successfully for termites by the age of 5 or 6 years. Before this, infants watched their mothers intently and "practiced" on their own, often showing abbreviated and interchanged elements. Another pattern of elementary technology, ant dipping, involves no play, and the mastery of skills occurs later. Also observed was leaf sponging for water from the tree holes, which becomes important during the dry season.

Goodall (1986) summarized data on Gombe chimpanzees' use of, and play with, various tools. Many of these tools were clearly used in playful acts, such as self-tickling or passing time by throwing stones in the air and trying to catch them. Tug-of-war games, using various pieces of vegetation, were also seen between young chimpanzees. To perfect the skill of termite fishing, a youngster poked at his mother's leg; to practice water foraging, he sponged at the base of a tree to extract water. All of these cases show the typical characteristics of object play: innovation, repetition, fragmentation, or substitution.

Another example of play in the development of elementary technology comes from Matsusaka and Kutsukake (2002), who described leaf sponging and leaf spooning by two juvenile female chimpanzees from M-group. Although leaf sponging (i.e., using a crushed leaf as a sponge) is not a common practice among Mahale's chimpanzees, these juveniles did it proficiently. Matsusaka and Kutsukake observed that the leaf spooning seemed to be more playful than functional in nature, because they could easily have gotten water to drink without using a leaf.

Tool use—using leaves to drink water—was also studied at Bossou by Tonooka (2001) in a group of 18 wild chimpanzees. Nine immature and nine adult chimps were studied for 33 days in the rain forest. The older infants versus adults and adolescents showed major differences in drinking with folded leaf cups; only certain types of leaves were used in this behavioral pattern. The functional act seemed to relate to object play. The report lacks details of individual performances but gives a flow chart of the tool-use sequence that allows the reader to compare the different paths for the adults and juveniles.

The first direct comparison of object play across field sites is provided by Nishida and Wallauer (2003), who monitored the chimpanzees of M-group in the Mahale Mountains National Park and the Kasakela community at Gombe, Tanzania, for a pattern of locomotor play with leaves, called leaf-pile pulling (LPL). The individual walks backward down a slope, facing a traveling party of chimpanzees, and rakes dry leaves on the ground with both hands, behind itself. This behavioral pattern occurs most often when a large party is in procession; the mother is often the recipient of the LPL performance by an infant or juvenile. According to Nishida and Wallauer, the pleasure in the process for the performer seems to come from the noise made by the dry leaves. Seventy-three bouts of LPL were observed: 11 by infant males, 22 by juvenile males, 6 by adolescent males, 3 by mature males, 11 by infant females, 19 by juvenile females, and one by an adolescent female. Adult females were never seen engaging in LPL. Nishida and Wallauer found that the LPL bouts of younger individuals were shorter in duration than those of older individuals and that 3- to 6-year-olds covered shorter distances than those ages 7 and older. For the 60 times in which the context of LPL was recorded, 39 occurred while traveling, 13 during stationary play, and 8 while resting.

Sometimes, anecdotes can alert us to more uncommon aspects of object play: Matsuzawa (1997) saw an 8-year-old female chimpanzee carry a small log of 50-centimeter length and 10-centimer diameter at the rainforest site of Bossou in Guinea. She carried and cradled the log as if it were a doll, much as do the local girls in Bossou. She solicitously treated the log as her mother had treated her younger sister when she was sick.

In another unusual example, Hirata, Yamakoshi, Fujita, Ohashi, and Matsuzawa (2001) twice watched young chimpanzees at Bossou interact with a hyrax (*Dendrohyrax dorsalis*) but without treating it as prey. In the first case, after the hyrax fell from a tree, a young male chimpanzee flailed it with an attached sapling, pausing occasionally, then suddenly left it and rejoined his group. The second case was of an 8-year-old male who beat a hyrax against a tree and swung it in the air, all while displaying a play face. A second chimpanzee, a female, later retrieved the now-dead hyrax, carried it around with her for the rest of the day and ultimately kept it overnight in her nest.

In another case study, Zamma (2002) reported an adult female chimpanzee having a rare interaction with a squirrel in the Mahale Mountains National Park in western Tanzania. Nkombo swung, shook, and dragged about a living squirrel. Displaying a play face, she manipulated the squirrel, while it tried to bite its attacker. Nkombo's playful mannerisms ended only when the squirrel died and her attempts to play with it yielded no response.

The only published account of object play in wild bonobos is Ingmanson's (1996) report of findings on tool use at the research site of Wamba, in the Democratic Republic of Congo (then Zaire). Although records of tool use are rare for wild bonobos, this researcher listed several examples, none of which was foraging, in contrast to chimpanzees. Solitary play with objects was the most frequent, comprising 55% of play bouts, while the other 45% were social play with objects. After the age of 3 years, object play was more often social than solitary, because by then infants had become more adept at independent locomotion as well as playing with peers. Sticks and small leafy branches were the main items used in play, with sticks often enhancing the playful interaction.

GORILLAS AND ORANGUTANS

Object-related behavior has been studied less often in gorillas and orangutans than in the genus *Pan*, perhaps because wild gorillas show no tool use, and wild orangutans have revealed their skills in elementary technology only recently.

Gorillas may use tools in captivity, however: Fontaine, Moissou, and Wickings (1996) observed 11 western lowland gorillas (*Gorilla gorilla gorilla*), in captivity in a research center in Gabon, seeking tool use. The focus of their study was mainly maternal behavior and infant development, but they noted all tool-use episodes. The use of sponges, ladders made from logs, and sticks as tools was observed, although only occasionally. The females were more skilled in tool use than the males, and tool use that led to food reward was shown by all but one member of the group.

Gorillas may also incorporate objects into "make-believe" play: Matevia, Patterson, and Hillix (2002) reported on the concept of pretense in gorillas (see also Gómez & Martín-Andrade, Chapter 7, this volume), which includes actions such as object substitution, in which an unusual object is used instead of another object, and animation, in which an inanimate object such as a doll is treated as an animate being. Their subject, Koko, a female western lowland gorilla, has been using sign language for almost three decades and is said to implement the notion of pretense in many different contexts. Whether it is a doll or a toy alligator, Koko plays with it much as a human child would do by feeding, cradling, or dressing it. In one instance, having been given a piece of gum, Koko used it, in turn, as a tooth filling, food for her toy lobster, lipstick, nasal mucous, and modeling clay. She also pretends to eat out of an empty bowl and to drink from an empty cup while making slurping sounds.

In a similar study, Gómez and Martín-Andrade (2002) discussed the pretend play habits of five wild-born captive gorillas, ages 2.5–3 years, housed at the Madrid Zoo. Categories of imaginative play, such as object substitution, context substitution, imaginary properties, and role enactment, were seen spontaneously in the gorillas. Two of the female infants handled stones, pieces of cloth, balls, and a doll in ways that resembled maternal behavior. They carried the items on the neck, under the arm and on the back, but no play faces were observed. One of the females once played a highly structured game of catch with a tennis ball, but this lasted only briefly. A few months later, she began the game again, this time using an apple. After playing catch with the apple, she replaced the old item for a new one, a piece of straw. According to the authors, the essence of the game was substituting one throwable object for another. Other pretend behavioral patterns included drinking from empty cups and pretending to construct a nest when inviting another to play.

Studies of the only Asian species of great ape (*Pongo pygmaeus*) have focused on both wild and rehabilitated orangutans. In the wild, van Schaik and colleagues (2003) surveyed six populations of wild orangutans (*Pongo pygmaeus*) in Borneo and Sumatra to compare cultural variants. The populations included four from Borneo: Gunung Palung, Tanjung Puting, Kutai, Lower Kinabatangan, as well as Ketambe and Suaq Balimbing in Sumatra. Some behavioral variants could be considered to be object play, such as snag riding, construction of play nests, and holding a leaf bundle while sleeping. Snag riding is absent in all but the Tanjung Putting population, where it is customary. Building nests for social play is customary in Gunung Palung, Tanjung Puting, and Leuser Ketambe, habitual in Leuser Suaq Balimbing, absent in Lower Kinabatangan, and present (with unknown frequency) in Kutai. The orangutan's rare holding of a leaf bundle while sleeping was likened to a doll by the authors. Four of the six populations do not show this behavioral pattern, but it occurs occasionally in the Gunung Palung and Tanjung Puting.

Russon, Vasey, and Gauthier (2002) reported on eye-covering play in released orangutans in Borneo, and compared this behavior to that of Japanese macaques (*Macaca fuscata*). The behavior was more common among orangutans than Japanese macaques; the Japanese macaques did not use tools for this activity, but the orangutans did once. Two adolescent orangutans, one male and one female, incorporated a shirt into their social game. They took turns placing shirts over the other's head and eyes, and the seeing companion charged and wrestled the "blind" partner. They continued wrestling until the blind partner pulled the shirt over the other's head. This game continued for over 15 minutes.

The review of published literature above shows that there are few studies that focus on object play, per se. Several studies of tool use and object manipulation exist, but they mostly omit to mention if juveniles were learning a skill by using object play. Data on object play by orangutans, bonobos, and gorillas are scarce, and studies of the four great apes in nature are less common than captive studies. The most plentiful information comes from object play in chimpanzees. This may well be a function of the rich array of tool-use behavior found in both wild and captive chimpanzees (McGrew, 1992).

Most of the published studies focus on immature apes, but comparisons between older juveniles and the younger infants lacked enough information to test hypotheses of ontogenetic processes. In the tightest comparison, between congeneric bonobo and chimpanzee, the results show that the latter uses more tools than the former, at least in the wild. No such comparisons here involved the gorilla. Captive bonobos, orangutans, and gorillas seem to be more apt to show object play than their wild counterparts, possibly because of the enforced leisure time available to a large-brained creature freed of the demands of foraging, migration, antipredator vigilance, and a complex social life. Thus, any comparison across taxa in captivity should control as much as possible for basic variables that may affect object play, such as sex, age, and rearing.

NORMS OF OBJECT PLAY IN WILD CHIMPANZEES

Object manipulation is not the same as object play. McGrew's (1977) study cited above gave baseline norms of object manipulation by infant and juvenile chimpanzees studied in nature in the Gombe National Park, Tanzania. All cases of hand or mouth contact with inanimate objects were recorded on a minute-by-minute basis. Results showed object manipulation to be common and consistent: It occurred, on average, in 75% of observation minutes (range = 69–76%, n = 6 subjects). The objects manipulated ranged from flowers to weaver ants' nests, in 25 categories, but overwhelmingly, vegetation was involved. Two-thirds of the time (range = 62–72%, n = 6 subjects) the objects were *in situ* trunk, branch, twig, leaf, flower, fruit, bark, and so on, growing as parts of tree, shrub, herb, grass, vine, and so on (Figure 5.1). Females directed a higher proportion of their manipulation to detached objects, that is, portable items that could be taken from place to place.

However, McGrew (1977) made no effort to separate playful from nonplayful object manipulation, nor did any other researcher cited above, except in single case studies that were not repeated. Here we present, for the

FIGURE 5.1. Two infant chimpanzees at Gombe (Prof., left; Gremlin, right) engage in social object play around an adult female (Melissa). Photograph by C. E. G. Tutin.

first time, analyses of systematically collected data on wild great apes in which playful object manipulation is separated from nonplayful manipulation or exploration. To do so, we extracted a subset of data, in which object manipulation co-occurred with either or both laughter or play face. Based on van Lawick-Goodall (1968), "laugh" (i.e., series of glottal, unclear expirations, usually with mouth wide open) and "play face" (i.e., mouth open, with minimal mouth retraction, usually showing only upper row of teeth) are unmistakable signals of playfulness.

Table 5.1 gives information on the five infant chimpanzees studied, ages 1.5–5 years. The two females and three males were studied over an average of 7 months, and were observable almost 90% of the time, between 07.30–18.00 hours. They averaged about 4,000 minutes of observation overall, in focal-subject sampling sessions that averaged about 150 minutes per session in duration. As strictly defined here, they averaged only 7.4% of the time in object play, but there was three-fold variation across subjects, from lows of 4.8% in Atlas and Gremlin to a high of 16.6% in Plato.

As seen in Table 5.2, play face was overwhelmingly more common overall than laughter (1,619 minutes vs. 142 minutes), and the ratios of these

TABLE 5.1. Infant Chimpanzee Subjects of Study at Gombe and Frequency of Object Play

Subject[a]	Sex[b]	Age[c]	Observability[d]	Months of observation[e]	Minutes of observation[f]	Minutes of object play	Minutes with object play (%)
Atlas	M	62–69	87	7	4,200	201	4.8
Skosha	F	33–40	93	7	3,960	293	7.4
Plato	M	27–30	87	3	2,340	389	16.6
Gremlin	F	25–29	90	4	3,600	173	4.8
Freud	M	18–25	89	7	6,000	705	11.8
Median		27–30	89	7	3,960	293	7.4

[a]For details on subjects, see Goodall (1986); here ranked by descending age.
[b]F, female; M, male.
[c]In months.
[d]Percentage of minutes in which subject was observable for majority of minute.
[e]Total months of study, between November 1973 and December 1974.
[f]See McGrew (1977, Table 1, p. 270).

TABLE 5.2. Frequency of Four Kinds of Object Play Showing Play Face and Laughter by Five Infant Chimpanzees

Category of manipulation[a]	Atlas		Skosha		Plato		Gremlin		Freud		Total	
	Pf	La	Pf	La	Pf	La	Pf	La	Pf	La	Pf	La
Handle	162	19	217	30	258	24	122	0	468	46	1,227	119
Mouth	11	0	29	4	79	8	34	0	147	7	300	19
Eat	5	1	8	1	20	0	12	0	19	0	64	2
Carry	2	1	4	0	0	0	5	0	17	1	28	2
Total	180	21	258	35	357	32	173	0	651	54	1,619	142

Note. Pf, Play face; La, Laugh.
[a]Handle, palmer surface of hand(s) contacts object; mouth, lips or mouth contacts object; eat, object inserted into mouth does not reemerge or, if it does, is altered; carry, object changes location without touching ground.

were highly correlated across subjects ($n = 5$, $r_s = .9$; $p = 0.05$). Infants showed more play faces and laughed more often, although Gremlin was never heard to laugh. The primary type of object manipulation in overall play was to handle it (76%, 1,346/1,761); handling was more common than the other three categories combined, across all subjects. The relative frequency of object play across subjects was the same across the four categories of manipulation. The ranked frequency from most to least was *handle, mouth, eat, carry.* Given this consistency, totals for play face and laughter and totals for the four types of play are combined for the remaining analyses.

Table 5.3 shows the top 10 categories or types of objects played with by the five infant apes. Apart from the consistent predominance of vegetation, there is little congruence in preference across subjects: from ranks 2–5, three subjects most preferred to play with sticks, but not Freud, who liked stones, or Atlas, who liked fruits. Of the 11 types of item listed, nine are natural objects but not cloth (which was pilfered from human accommodation or stolen from drying lines) or bananas (which were fed to chimpanzees as part of occasional provisioning). The non-natural components of the wild chimpanzee infants' toy-kit represent a trivial proportion, less than 0.5% of instances.

What can be said about object play by apes in nature and its implications for captive husbandry? Given an active day of about 12 hours, infant chimpanzees spend an hour or two a day in object play, as strictly defined

TABLE 5.3. Category of Object Played with by Five Infant Chimpanzees, Ranked by Total Frequency

Category of object	Atlas	Skosha	Plato	Gremlin	Freud	Total
Vegetation	145	207	252	87	461	1,152
Leaf	2	21	29	17	57	126
Stick	3	27	32	20	39	121
Fruit	10	6	22	12	26	76
Stone	7	4	5	2	41	59
Banana	0	14	2	5	12	33
Grass	9	3	4	13	1	30
Vine	6	2	10	0	7	25
Cloth	0	0	0	0	23	23
Earth	2	1	10	3	1	17
Miscellaneous	17	8	23	14	37	99
Total	201	283	389	173	705	1,761

here (Figure 5.2). This object play occurs in addition to non-object locomotor and social play, and to object manipulation and exploration that cannot, with certainty, be called playful. The natural world is full of play items, mostly vegetation, which when intact provides a three-dimensional substrate, or when detached provides portable toys. The fifth most-often used toys are stones; this finding is notable because some populations of chimpanzees show customary lithic technology; for example, stones used as hammers and anvils to crack open nuts (Boesch & Boesch, 1983; Sugiyama & Koman, 1979). Gombe's chimpanzees do not engage in this particular behavior pattern, but it cannot be due to lack of familiarity with stones, as shown here (see also McGrew et al., 1997).

Gombe's chimpanzees customarily make and use tools made of detached vegetation: leaves, to "sponge" for water from tree holes; sticks, to dip for driver ants; and grass blades and vines, to fish for termites (Whiten et al., 2001). More often than any of these other types of elementary technology, however, they daily make shelter (nests or beds) from intact vegetation. Each wild chimpanzee past the age of weaning makes at least one such platform in the trees for overnight sleeping, every day of its life (Goodall, 1962). This overarching reliance on functional object manipulation for daily shel-

FIGURE 5.2. Infant chimpanzee (Wilkie) at Gombe manipulates a palm frond with right hand while clinging to his mother (Winkle) with left hand. Photograph by C. E. G. Tutin.

ter and for occasional subsistence through extractive foraging is reflected in the object play of the youngsters.

Finally, we can consider the implications of these findings for apes held captive in laboratories or zoological parks. Few have access to growing woody vegetation; when they do, it is "hot-wired" to keep them from touching it. The reason for this deprivation of natural raw materials is straightforward: Most captive apes live at such high density that the vegetation could not recover fast enough from their destructive attention, were they given free access. Thus, it is not surprising that captive-born and -reared chimpanzees are poor nest builders (Videan, 2003). The next best alternative is to give them lots of "browse" (cut leafy and woody vegetation) and hay or straw. Good zoos do this, but most labs cannot be bothered, as such materials clog drains and add to cleaning chores. Finally, no establishment housing captive apes provides them with stones—just the opposite: These are removed, lest they be used as missiles! Overall, by wild standards, it is clear that object play is normal, at least for young apes, and any captive facility that withholds the raw materials for object play is depriving them of a basic component of what should be a nurturing environment.

What is not clear from this limited study of a few wild chimpanzees in Tanzania is whether or not the findings generalize to greater numbers of the species or to other species of apes. Given what we know about behavioral diversity within chimpanzees (Whiten et al., 2001) and across great apes (Hohmann & Fruth, 2003; van Schaik et al., 2003), this generalizability cannot be assumed without systematic, comparative study. It may be that access to a variety of objects is more important to species of great apes that uses tools customarily (chimpanzee, orangutan) than to those that do not (bonobo, gorilla).

OBJECT PLAY COMPARED IN CAPTIVE CHIMPANZEES AND GORILLAS

Accordingly, we designed a study comparing the two genera (*Pan* and *Gorilla*) of African great apes, focusing on their playful responses to a standard set of introduced objects. To make the comparison even tighter, we sought to match the more limited set of gorillas available in the zoo with the larger population of chimpanzees. Thus, the aim was to test for species differences, while controlling for the variables of sex, age, and rearing experience.

Table 5.4 gives details of the subjects: All 15 gorillas were matched for sex; the median age difference within a matched pair was 2 years; and 14 of 15 pairs were matched as either wild or captive reared. (Within the category

TABLE 5.4. Subjects Matched across Species by Sex, Age, and Rearing

Gorilla (n = 15)	Sex	Age	Rearing[a]	Chimpanzee (n = 17)	Sex	Age	Rearing
Colo	F	46.5	Hand	Mae	F	38	Wild
Pongi	F	40	Wild	Abbey	F	38	Wild
Sylvia	F	38	Wild	April	F	23	Wild
Lulu	F	39	Wild	Junie	F	37	Wild
Mumbah	M	38	Wild	Pacer	M	32	Wild
Nia	F	9.5	Hand	Hannah	F	12.5	Mother
Nkosi	M	12	Hand	Marcus	M	11.5	Mother
Jontu	M	6.5	Mother	Rusty	M	7.5	Mother
Macombo II	M	19.5	Hand	Spider	M	16.5	Mother
Jumoke	F	13.5	Hand	Tina	F	17.5	Mother
Kebi Moyo	F	12.5	Mother	Sophie	F	15	Mother
Kambera Dupe	F	4.5	Hand	Lexus	F	5	Mother
Muchana	M	3	Mother	Gage	M	2	Mother
Cassie	F	10	Mother	Cassie	F	14	Mother
Little Joe	M	5.5	Hand	Lyle	M	4.5	Mother
				Sindee	F	4	Mother
				Alpha	F	18.5	Mother
	F = 9 M = 6	Median = 12.5	Wild = 4 Captive = 11		F = 11 M = 6	Median = 15	Wild = 5 Captive = 12

[a]Hand, hand reared by humans; mother, mother reared in captivity; wild, born in wild and mother reared.

of captive rearing, however, there was a difference between the species: the zoo gorillas were more often hand reared, whereas the lab chimpanzees were all mother reared.) Two chimpanzees (Alpha and Sindee) were added to the chimpanzee sample, because they were members of the groups being observed, and so interacted with the subjects and the objects.

We hypothesized that, if play is practice for functional tool use, then a tool-using species would be more inclined to benefit from such a practice than a non-tool-using species. In nature, tool use is ubiquitous for chimpanzees; it is found in more than 40 wild populations across Africa (McGrew, 1992; Whiten et al., 1999). In contrast, wild gorillas have never been seen to use tools (see review in McGrew, 1992). Thus, we predicted that in a tightly controlled comparison such as this, chimpanzees would show more object play than gorillas, even in captivity.

Seventeen chimpanzees were observed in three multimale, multifemale groups at the University of Texas MD Anderson Cancer Center, Bastrop, Texas. Observations were done 5 or 6 days a week, from May 20 to June 20, 2003. The 10 females and 7 males chosen as subjects (see below) lived in outside corrals with other chimpanzees (not included in the study). Fifteen western lowland gorillas were also observed in two groups at the Columbus Zoo and Aquarium, Columbus, Ohio. The observations were carried out 5 days a week, from June 30 to July 25, 2003. The chimpanzees were chosen to match the age and sex of the nine female and six male gorillas housed at the Columbus Zoo (Figures 5.3 and 5.4).

Groups of chimpanzees or gorillas were given two types of items, one artificial and one natural, in order to promote object play: rubber tubs and stems of bamboo. Only one type of object was in the enclosure at any one time, and these were given on a weekly basis, on about 5 days for each object for each group. At the onset of passing out the objects, a 1-hour scan sample was done on the group being provisioned. In addition to the scan sampling, 10 to 12 of 10-minute focal-subject samples were taken on each of the 32 individuals. Object play was coded for the same four categories as used in the Gombe study (see above): handle, mouth, carry, and eat. Play faces and bouts of playful behavioral patterns were noted as the true frequencies.

FIGURE 5.3. Juvenile male chimpanzee (Lyle) at the University of Texas, M. D. Anderson Cancer Center, demonstrates solitary object play with a large rubber tub. Photograph by J. K. Ramsey.

FIGURE 5.4. Adolescent female gorilla (Nia) at Columbus Zoo playfully nests in large rubber tub. Photograph by J. K. Ramsey.

Three large (approximately 61 centimeters diameter, 24 centimeters deep) and three small (approximately 25 centimeters diameter, 8.5 centimeters deep) black rubber tubs were added to each enclosure. For the chimpanzees, the tubs were given daily and allowed to accumulate over the week because of the inconvenience of going into the enclosure every day to change them and the strain such an activity would cause in the subjects. At the end of the week, the tubs were removed. Because the two groups of gorillas time-shared the outdoor exhibit enclosure daily at the zoo, the tubs were left in the enclosure and taken out daily for a week. Each chimpanzee and gorilla individual was given daily fresh-cut stems of bamboo, 45–60 centimeters long, for 5 days in a row. Other objects were already in the enclosures: ropes, balls, sticks, grass, nesting material (woodwool, newspaper, cardboard, hay) pool of water (gorilla) or puddle of water (chimpanzee).

RESULTS

Table 5.5 gives the results of the comparison of the chimpanzees and gorillas, by type of object, with all four manipulations combined. Chimpanzees averaged five times more frequent play with the rubber tubs than did gorillas, a significant difference. For the bamboo browse, the mean difference was

TABLE 5.5. Mean Frequency of Playful Manipulation of Objects by Captive Chimpanzees and Gorillas

	Stimulus object			
Species	Tub	Bamboo	Other	Total
Chimpanzee (n = 17)	7.0	2.8	2.4	12.5
Gorilla (n = 15)	1.4	0.5	2.5	4.8
z score[a]	2.15	1.28	0.16	0.41
p	0.03	0.10	0.44	0.34

[a]Mann–Whitney U test, one-tailed.

almost six-fold in favor of the chimpanzees, but the result was only a nonsignificant trend. There was no difference between species for play with other familiar items. The overall measure of all object play combined showed an apparent difference in favor of chimpanzees, but the amount was not statistically significant.

As expected, the chimpanzees played more and differently with the tubs and browse than did the gorillas. The younger chimpanzees engaged in little nest-building play with the browse but more games of "keep away." However, in most cases, the chimpanzees as well as the gorillas ate the browse as soon as they received it. The introduction of the tubs, however, drew a large amount of attention from both species. The tubs were completely unknown to most of the chimpanzees, unlike the gorillas, which had received them regularly until 2 months prior to the study. The younger chimpanzees mainly used the tubs as a new element on which to perform cartwheels and flips, as well as hiding underneath, much like a turtle shell. In one instance, as the week progressed and more tubs accumulated, Hannah played at nest building with seven or eight of the small and large tubs. The gorillas made use of the pool in their enclosure by dunking or standing on the tubs and splashing around. In several instances, Nkosi used a small tub to splash around the water, with or without others nearby.

Comparison of the chimpanzees' and gorillas' play with the novel tubs was in the predicted direction, as were the near-significant results with the more familiar browse. Thus, the tool-using species predominated. The ever-present other objects were given equal attention by the two species, perhaps because of the objects' low salience due to overexposure or boredom. In any event, the idea that play provides practice for more functional object manipulation in the form of tool use is supported, but a direct study of tool use in the same subjects would be required to confirm this hypothesis.

DISCUSSION

Previous studies of object play, both in nature and captivity, have failed to separate play from the broader category of object manipulation. A review of the published literature on the topic in great apes shows that such distinctions are often ignored, or if acknowledged, are not addressed. Here, by focusing on object manipulation accompanied by the unequivocal signal of the play face or laughter (Loizos, 1967, 1969), we tackled this omission. The first study, of object play in the wild chimpanzees, sought to establish species norms in the natural habitat. Its strengths were its intensive data gathered over 40–100 hours of observation per subject, and over a wide variety of manipulated objects. Its weaknesses were the few ($n = 5$) subjects, all of whom were infants or juveniles, so it should be treated as a pilot study. The results show a reassuring congruence between object play by youngsters and the elementary technology characteristic of the population (e.g., Goodall, 1962).

The second study, of two genera of apes in captive settings, aimed to show that ecologically valid phenomena can be examined under artificial conditions. Most of the captive-born subjects (23 of 32) had never seen an African forest and so could not be aware of the species difference between *Pan* and *Gorilla* in the natural occurrence of elementary technology. Nor had the nine wild-born ape subjects been given the chance in captivity to fish for termites, sponge for water, crack nuts, and so forth, and so to act as models for their captive-born counterparts. Thus, the differences found in the predicted direction are all the more remarkable for their spontaneity in surroundings both enriched (by captive standards) but impoverished (by natural standards).

It is sometimes assumed that everything important is known about object play in great apes, at least in the well-studied chimpanzee, but this is not so. Both normative and comparative studies of the other three species of great ape—bonobo, gorilla, and orangutan—are lacking, with a few exceptions (Takeshita & Walraven, 1996). This paucity may be understandable for the bonobo, which is rare in captivity (fewer than 100 individuals worldwide) and barely studied in nature (indigenous to war-torn Congo). It is not understandable for captive gorilla and orangutan, which can be found in the hundreds in zoos, if not in laboratories. Neither species is easily studied in nature, but there are habituated populations of both, under behavioral observation.

There are at least six sites in Africa (Bossou, Budongo, Gombe, Kanyawara, Ngogo, Tai) where close-range behavioral data can be collected on wild chimpanzee subjects. Much is known of their tool use but little of their object play (McGrew, 1977). There are no hypothesis-driven studies of object play in wild chimpanzees, despite more than 40 years of field study since Jane Goodall began her work in 1960.

More is known about chimpanzees in captivity, which is not surprising, given that the worldwide captive population of chimpanzees probably numbers between 3,000–4,000, in zoological gardens, safari parks, refuges and sanctuaries, and laboratories. There are huge databases available, such as Chimpanzoo, a collective effort of 12 zoos and coordinated by the Jane Goodall Institute, and some laboratories have scores or even hundreds of chimpanzees on a single site. Yet, little has emerged by way of published research to elucidate object play by the species. This lack of knowledge hampers the comparison of humans and their nearest living relations, as noted in this volume. If play is essential to normal human development, and if normal human development is rooted in the evolutionary past of our species, and if we can only hope to infer that evolution from comparative studies of humanity's nearest living relations, then more directly linked research needs to be done by primatologists and psychologists.

Finally, there is an ethical reason for studying object play in apes: We need to know norms from nature in order to provide suitable environmental enrichment for their captive counterparts. If wild apes have material culture, and if we must keep apes in captivity, then at least we must try to give the captive apes enough varied stimulation to make up for what they are missing.

ACKNOWLEDGMENTS

We thank Dr. Steve Schapiro, Susan Lambeth and the staff at University of Texas MD Anderson Cancer Center, Beth Pohl and the African Forest staff at Columbus Zoo, and Dr. Linda Marchant for their encouragement, support, and cooperation. The research was funded by a Miami University Undergraduate Summer Scholarship to Jacklyn K. Ramsey.

REFERENCES

Adang, O. M. J. (1984). Teasing in young chimpanzees. *Behaviour, 88*, 98–122.

Beck, B. B. (1980). *Animal tool behavior: The use and manufacture of tools by animals.* New York: Garland.

Bjorklund, D. F., Yunger, J. L., Bering, J. M., & Ragan, P. (2002). The generalizations of deferred imitation in enculturated chimpanzees (*Pan troglodytes*). *Animal Cognition, 5*, 49–58.

Boesch, C., & Boesch, H. (1983). Optimisation of nut-cracking with natural hammers by wild chimpanzees. *Behaviour, 83*, 265–286.

Burghardt, G. M. (1998). The evolutionary origins of play revisited: Lessons from turtles. In M. Bekoff & J. A. Byers (Eds.), *Animal play: Evolutionary, comparative, and ecological perspectives* (pp. 1–26). Cambridge, UK: Cambridge University Press.

Fagen, R. (1981). *Animal play behavior.* Oxford, UK: Oxford University Press.

Fagen, R. (1993). Primate juveniles and primate play. In M. E. Pereira & L. A. Fairbanks (Eds.), *Juvenile primates: Life history, development, and behavior* (pp. 182–196). Oxford, UK: Oxford University Press.

Fontaine, B., Moisson, P. Y., & Wickings, E. J. (1996). Observations of spontaneous tool making and tool use in a captive group of western lowland gorillas (*Gorilla gorilla gorilla*). *Folia Primatologica, 65*, 219–223.

Gómez, J. C., & Martín-Andrade, B. (2002). Possible precursors of pretend play in nonpretend actions of captive gorillas (*Gorilla gorilla*). In R. W. Mitchell (Ed.), *Pretending and imagination in animals and children* (pp. 255–268). Cambridge, UK: Cambridge University Press.

Goodall, J. (1962). Nest building behavior in the free ranging chimpanzee. *Annals of the New York Academy of Sciences, 102*, 455–467.

Goodall, J. (1986). *The chimpanzees of Gombe: Patterns of behavior.* Cambridge, MA: Harvard University Press.

Hall, S. L. (1998). Object play by adult animals. In M. Bekoff & J. A. Byers (Eds.), *Animal play: Evolutionary, comparative, and ecological perspectives* (pp. 45–60). Cambridge, UK: Cambridge University Press.

Hayes, C. (1951). *The ape in our house.* New York: Harper & Row.

Hirata, S., Yamakoshi, G., Fujita, S., Ohashi, G., & Matsuzawa, T. (2001). Capturing and toying with hyraxes (*Dendrohyrax dorsalis*) by wild chimpanzees (*Pan troglodytes*) at Bossou, Guinea. *American Journal of Primatology, 55*, 93–97.

Hohmann, G., & Fruth, B. (2003). Culture in bonobos? Between-species and within-species variation in behavior. *Current Anthropology, 44*, 563–571.

Ingmanson, E. J. (1996). Tool-using behavior in wild *Pan paniscus.* In A. E. Russon, K. A. Bard, & S. T. Parker (Eds.), *Reaching into thought: The minds of the great apes* (pp. 191–210). Cambridge, UK: Cambridge University Press.

Kellogg, W. N., & Kellogg, L. A. (1933). *The ape and the child.* New York: McGraw-Hill.

Ladygina-Kohts, N. N. (2002). *Infant chimpanzee and human child.* Oxford, UK: Oxford University Press.

Lethmate, J. (1982). Tool using skills in orangutans. *Journal of Human Evolution, 11*, 49–64.

Loizos, C. (1967). Play behaviour in higher primates: A review. In D. Morris (Ed.), *Primate ethology* (pp. 176–218). Chicago: Aldine.

Loizos, C. (1969). An ethological study of chimpanzee play. In C. R. Carpenter (Ed.), *Proceedings of the second international congress of primatology* (Vol. 1, pp. 87–93). Zurich: S. Karger.

Markus, N., & Croft, D. B. (1995). Play behaviour and its effects on social development of common chimpanzees (*Pan troglodytes*). *Primates, 36*, 213–225.

Matevia, M. L., Patterson, F. G., & Hillix, W. A. (2002). Pretend play in a signing gorilla. In R. W. Mitchell (Ed.), *Pretending and imagination in animals and children* (pp. 285–304). Cambridge, UK: Cambridge University Press.

Matsusaka, T., & Kutsukake, N. (2002). Use of leaf-sponge and leaf-spoon by juvenile chimpanzees at Mahale. *Pan African News, 9*, 6–9.

Matsuzawa, T. (1997). The death of an infant chimpanzee at Bossou, Guinea. *Pan African News, 4*, 4–6.

McGrew, W. C. (1977). Socialization and object manipulation of wild chimpanzees. In S. Chevalier-Skolnikoff & F. E. Poirier (Eds.), *Primate bio-social development: Biological, social, and ecological determinants* (pp. 261–288). New York: Garland.

McGrew, W. C. (1992). *Chimpanzee material culture: Implications for human evolution.* Cambridge, UK: Cambridge University Press.

Mendoza-Granados, D., & Sommer, V. (1995). Play in chimpanzees of the Arnhem Zoo: Self-serving compromises. *Primates, 36*, 57–68.

Nishida, T., & Wallauer, W. (2003). Leaf-pile pulling: A suspected tradition among the chimpanzees of Mahale. *American Journal of Primatology, 60*, 167–173.

Pellegrini, A. D., & Smith, P. K. (2002). Children's play: A developmental and evolutionary orientation. In J. Valsiner & K. Connolly (Eds.), *Handbook of developmental psychology* (pp. 276–291). London: Sage.

Poti, P., & Spinozzi, G. (1994). Early sensorimotor development in chimpanzees (*Pan troglodytes*). *Journal of Comparative Psychology, 108*, 93–103.

Power, T. G. (2000). *Play and exploration in children and animals.* Mahwah, NJ: Erlbaum.

Russon, A. E., Vasey, P. L., & Gauthier, C. (2002). Seeing with the mind's eye: Eye-covering play in orangutans and Japanese macaques. In R. W. Mitchell (Ed.), *Pretending and imagination in animals and children* (pp. 241–254). Cambridge, UK: Cambridge University Press.

Sugiyama, Y., & Koman, J. (1979). Tool-using and making behavior in wild chimpanzees at Bossou, Guinea. *Primates, 20*, 513–524.

Takeshita, H., & Walraven, V. (1996). A comparative study of the variety and complexity of object manipulation in captive chimpanzees (*Pan troglodytes*) and bonobos (*Pan paniscus*). *Primates, 37*, 423–441.

Temerlin, M. K. (1975). *Lucy: Growing up human—a chimpanzee daughter in a psychotherapist's family.* Palo Alto, CA: Science and Behavior Books.

Tonooka, R. (2001). Leaf-folding behavior for drinking water by wild chimpanzees (*Pan troglodytes verus*) at Bossou, Guinea. *Animal Cognition, 4*, 325–334.

van Lawick-Goodall, J. (1968). The behaviour of free-living chimpanzees in the Gombe Stream Reserve. *Animal Behaviour Monographs, 1*, 161–311.

van Schaik, C. P., Ancrenaz, M., Bergen, G., Galdikas, B., Knott, C. D., Singleton, I., Suzuki, A., Utami, S. S., & Merrill, M. (2003). Orangutan cultures and the evolution of material culture. *Science, 299*, 102–105.

Vauclair, J. (1984). Phylogenetic approach to object manipulation in human and ape infants. *Human Development, 27*, 321–328.

Videan, E. N. (2003). [Sleep and nest-building in captive chimpanzees]. Unpublished raw data.

Whiten, A., Goodall, J., McGrew, W. C., Nishida, T., Reynolds, V., Sugiyama, Y., Tutin, C. E. G., Wrangham, R. W., & Boesch, C. (1999). Cultures in chimpanzees. *Nature, 399*, 682–685.

Whiten, A., Goodall, J., McGrew, W. C., Nishida, T., Reynolds, V., Sugiyama, Y., Tutin, C. E. G., Wrangham, R. W., & Boesch, C. (2001). Charting cultural variation in chimpanzees. *Behaviour, 138*, 1481–1516.

Yerkes, R. M. (1943). *Chimpanzees. A laboratory colony.* New Haven, CT: Yale University Press.

Zamma, K. (2002). A chimpanzee trifling with a squirrel: Pleasure derived from teasing? *Pan African News, 9*, 9–11.

CHAPTER 6

Boys' and Girls' Uses of Objects for Exploration, Play, and Tools in Early Childhood

ANTHONY D. PELLEGRINI AND KATHY GUSTAFSON

In this chapter we describe different ways in which preschool-age children interact and play with objects. We differentiate "play" with objects from other sorts of object use, such as exploration, construction, and tool use. This distinction is an important one because each seems conceptually distinct and has different developmental trajectories and possibly different functions.

OVERVIEW OF THE LITERATURE

We know from the early work of Hutt (1966) that exploration of objects is an information-gathering venture. An object is manipulated manually, visually scanned, and sometimes tasted to find out "what it is." This knowledge is necessary but not sufficient for individuals to play with the objects. Consequently, exploration precedes play ontogenetically; exploration is observed from at least 7½ months of age and declines as play with objects begins and increases (Belsky & Most, 1981). Play, on the other hand, reflects an orientation that Hutt typified as "What can I do with it?" Thus, play is less instrumental and more means oriented than exploration, tool use, or construction. Play begins as exploration is decreasing and continues to increase during childhood (Belsky & Most, 1981) and then later decrease (Fein, 1981).

Construction is treated as play behavior by some (e.g., Rubin, Maioni, & Hornung, 1976), whereas others do not consider it to be play (e.g., Piaget, 1962; Smith, Takhvar, Gore, & Vollstedt, 1986). The developmental trajectory of construction certainly does not follow the traditional inverted-U developmental function of other forms of play. The development of construction abilities remains flat across much of the preschool period, though it accounts for a substantial portion of children's activity budget, around 40% (Rubin et al., 1976).

Tool use is more like construction than play, given its instrumental flavor, but it has not been studied extensively by developmental psychologists. Thus, we do not have information on its developmental trajectory or the amount of time allocated to tool use during childhood. In this chapter we describe the occurrence of different forms of object use during the preschool period. Our goal is to establish a time budget of different forms of object use and examine the degree to which each relates to children's cognitive ability to use objects to solve different sorts of problems.

The role of objects in children's development has received an especially prominent focus in the cognitive developmental literature. At a distal level, object manipulation has assumed an important role in some accounts of the evolution of human cognition (for a review, see Tomasello & Call, 1997). More proximally, theorists as different as Montessori (1964) and Piaget stressed the role of children's experiences with objects as integral to the development of cognitive processes. For example, Piaget's (1983) theory of sensorimotor intelligence has been particularly influential in assigning an important role to objects in the development of intelligence.

Indeed, much of the study of children's interactions with objects during childhood has been influenced directly by Piagetian theory. For example, Smilansky's (1968) influential monograph included "functional play" and "constructive play" as categories of children's interactions with objects. Functional play, which was similar to Piaget's (1962) notion of practice play, is characteristic of the sensorimotor period of development when children interact with objects to gain mastery. These non-goal-oriented actions with objects, according to Piaget (1962), result from children's pleasure in mastering action-based routines.

In "constructive play," according to Smilansky, the child learns the "various uses of play materials" and the "building" of something (1968, p. 6). Ironically, perhaps, much of what we know about children's interactions with objects is subsumed under research on constructive play. We say "ironic" because construction, according to Piaget (1962), was considered to be less playful than work-like. That is, construction is more concerned with the end product of the activity—the constructed object, per se—whereas play is more concerned with the activity, or means, than with the end product, per se (Piaget, 1962; Smith et al., 1986).

Rubin and colleagues (Rubin et al., 1976; Rubin, Watson, & Jambor, 1978) expanded Smilansky's categories of play into a heuristic for describing the social (solitary, parallel, and interactive) and cognitive (functional, constructive, and dramatic/pretense) dimensions of play. In this scheme, Rubin and colleagues considered constructive "play" to involve manipulation of objects to create something. This scheme has been used to generate massive amounts of descriptive data on the ways in which young children use objects (for a full review, see Rubin, Fein, & Vandenberg, 1983). However, Rubin and colleagues (Rubin et al., 1983) rightfully questioned the validity of "constructive play" as a category because of its incongruity with Piagetian theory: "Constructive play might be viewed as belonging to some other coding schemes" (p. 727).

These descriptive data are, at the same time, both valuable and limited. They are valuable to the extent that they provide some descriptions of children's object use but limited in that the category "constructive play" is probably too gross to offer guidance about the role of objects in children's cognitive development. The above definition given by Smilansky and subsequently used by Rubin (Rubin et al., 1976, 1978) includes a diverse constellation of goal-directed uses of objects. For example, using blocks to build steps might be considered constructive play. However, the same act, though coded as constructive, might actually be considered "tool use" if a child uses the steps to enhance his or her reach.

The loss of descriptive information that occurs by using the term constructive play to include interactions with objects has implications for our understanding of the development of children's uses of objects as well as the functions of those interactions with objects. That is, exploration of objects, play with objects, construction, and using objects as tools probably have different developmental histories (antecedents) as well as different outcomes (consequences). For example, constructive play, as defined in Rubin's adaptation of Smilansky's categories, does not show age-related changes (at least from the ages of 3–5 years) when children are observed in their preschool classrooms (Pellegrini & Perlmutter, 1989; Rubin et al., 1978). That constructive play does not follow the typical inverted-U developmental curve of play, as noted above, is further evidence that it probably has different antecedents and is not play in the strict sense that "play" is more means than ends oriented.

Exploration is different from both play and tool use (Hutt, 1966). Exploration is the behavior exhibited when individuals first encounter objects and explore their properties and attributes. Exploration shows developmental change both microgenetically, as it increases with the presentation of a new object and then decreases (Hutt, 1966), and with age, decreasing during childhood (Henderson & Moore, 1979). Regarding tool use, age trends also are evidenced, with skills increasing from infancy (Connolly & Dalgleish, 1989; Connolly & Elliot, 1972). Studies of tool use in childhood show increases in facility with age, but the findings are drawn almost solely from performances on experimental tasks, thus lacking ecological validity (e.g., Bates, Carlson-Luden, & Bretherton, 1980).

Developmental descriptions of children's use of objects in their everyday worlds, encompassing exploration, play, construction, and tool use, are sorely lacking (Power, 2000). We simply do not have descriptions of the time and effort that children spend using objects in the preschool or home environments. Time-budget descriptions of tool use, as well as other forms of object use, are important for two reasons. First, they would complement the age-related descriptions of toddlers' (Chen & Seigler, 2000) and children's tool use in contrived studies (Smitsman & Bongers, 2003; Vandenberg, 1980; van Leeuwen, Smitsman, & van Leeuwen, 1994). Second, such field-based descriptions of preschoolers' use of tools would be an important complement to Connolly's descriptions of infants' and toddlers' tool use (Connolly & Dalgleish, 1989; Connolly & Elliot, 1972).

Our descriptions of time spent in different types of object use during childhood are framed in terms of human behavioral ecology theory (e.g., Blurton Jones, Hawkes, & O'Connell, 1997). Specifically, descriptions of the "costs" (time spent in different forms of object use) associated with an activity serve as an indicator of its importance or function. High costs correspond to high benefits, and low costs, usually, correspond to low benefits (Caro, 1988). Costs are typically documented in terms of the resources (time and energy) expended to acquire or learn a skill. "Time in an activity" is typically expressed as the portion of the total time budget spent in that activity (Martin, 1982), and "energy" is typically expressed in terms of caloric expenditure in that activity relative to the entire caloric budget (Pellegrini, Horvat, & Huberty, 1998). In this chapter we document cost only in terms of time. By way of comparison, chimpanzees spend between 10% and 15% of their waking time in tool use (McGrew, 1981); thus, if the chimpanzee is a model for understanding human evolution (see below), we would expect similar levels in human juveniles.

The logic of this level of analysis is that learning and acquiring specific skills involve different tradeoffs between costs and benefits, and individuals tend to adopt the most "efficient," or optimal, strategies (Boyd & Richerson, 1985). For example, in learning to use tools, tradeoffs must be made between different methods in light of finite time and energy budgets. Individuals' decisions tend to be optimal to the extent that they generally choose the most efficient strategy (Boyd & Richerson, 1985). From this view there should be a correspondence between time budgets and the benefits associated with expenditures in each activity.

This level of analysis has proven to be useful in documenting the role of certain forms of play in children's development. For example, caloric costs of children's locomotor play tends to be "moderate," accounting for about 5% of the time and energy available during childhood (Pellegrini et al., 1998), a figure consistent with the comparative literature (Fagen, 1981). The moderate costs correspond to moderate benefits of play, again consistent with the correlational and experimental data in child and educational psychology (Martin & Caro, 1985).

Time spent in different forms of object use should predict children's ability to use and construct tools to solve divergent and convergent problems. However, the experimental studies examining the relation between interaction with objects and subsequent use of objects to solve problems have employed very brief periods of object interaction, usually about 10 minutes (Dansky & Silverman, 1973; Smith & Simon, 1984; Sylva, Bruner, & Genova, 1976). A more valid test of function would be to regress time spent in different forms of object play, including tool use, on to tool use facility.

Making functional inferences about the development of behavior and skills based on cost:benefit ratios has been effectively employed by Bock (1999; Chapter 10, this volume) in the area of object play and tool use, in his ethnographic work among the Okavango Delta people. Specifically, using an extensive corpus of direct behavioral observation, he described the development of different forms of object play and tool use for boys and girls from birth through 18 years of age. Playful tool use took the form of play pounding, wherein children, usually girls, take a stick, reed, or mortar and pound dirt. This activity is similar to the grain-processing work in which women engage. Indeed, groups of young girls often initiated this task close to adult women actually engaging in the task. Bock also designed an experimental task in which girls' grain-processing efficiency was assessed.

Bock's results indicated that boys engaged in more object play than girls. Additionally, boys' object play followed an inverted-U curve, typical of the development of play (Fagen, 1981), accounting for 0.11 of the observations at 0–3 years, 0.17 at 4–6, and 0.03 at 7–9 years. For girls, an inverted-U curve was also observed for object play, but the peak was at a later age:

0.04 at 0–3 years, 0.08 at 4–6 years, 0.11 at 7–9 years, 0.15 at 10–12 years, and zero at 13–15 years. The trajectory for play pounding by girls paralleled object play: 0.05, 0.16, 0.17, 0.22, and zero, respectively.

It is important to note the similarity between these time budgets and those documented by McGrew (1981) for chimpanzees. Additionally, girls spent more time in play pounding than did boys. For girls, time spent in play pounding predicted grain-processing facility in an analogue task. Thus, skills learned and practiced in relatively unstructured regimens, such as object play, predicted later facility with using objects as tools in a productive task.

Correspondingly, ethnographic and ethological studies of tool use in human and nonhuman primates have been conducted with a particular eye on sex differences (McGrew, 1992; Power, 2000). This level of description is similar to that proffered by McGrew (1981; Ramsey & McGrew, Chapter 5, this volume) for male and female chimpanzees, *Pan troglodytes.* Chimpanzees are an especially important species for human comparisons because they are genetically our closest relatives, and we share many social and cognitive features (McGrew, 1981; Tanner, 1987; Wrangham, 1987). Comparing the ontogenetic development of humans' tool use with that of chimpanzees is an important step in understanding the meaning and function of those behaviors (Tinbergen, 1963). Specific to chimpanzees' tool use, McGrew suggests that tool use precedes tool construction and that female chimpanzees use tools primarily for gathering and processing food, often in sedentary contexts.

A paradigmatic example of this sort of activity involves females using sticks or stalks of grass to fish for termites. Males, on the other hand, use tools as weapons in hunting and predation, in a vigorous manner. For example, a stick can be used against another chimpanzee, or a stone can be thrown at prey. Thus, males and females both use objects and tools in sophisticated but very different ways. For both sexes, tool use occupies about 10–15% of their waking time budget.

If there were a female bias in using and making tools for gathering, we would expect a corresponding sex difference that favors females in some forms of tool use, such as raking and digging. Furthermore, whereas female chimpanzees use objects for gathering, male chimpanzees use objects and tools as weapons, in the service of hunting (McGrew, 1992).

These predictions are supported in the human foraging literature (see Gosso, Otta, Morais, Ribeiro, & Raad Bussab, Chapter 9, this volume). For example, forager boys (in the Amazon), more than girls, are frequently seen playing with bows and arrows and sling-shots. Similarly, the forager literature supports the idea that girls make tools associated with gathering more frequently than boys. For example, making baskets out of palm leaves is a common form of play among the Parakanã girls from Pará State (see Gosso et al., Chapter 9, this volume). Indeed, this activity was observed exclusively among girls.

Describing the various forms of object use should also help clarify inconsistent and confusing sex difference data in the child developmental literature. For example, in most of the studies using the Smilansky–Parten matrix, females are reported to engage in constructive play more than males (Johnson & Erlsher, 1981; Johnson, Erlsher, & Bell, 1980; Rubin et al., 1976, 1978). However, boys' constructions tend to be more complex than girls' (Erickson, 1977; see Rubin et al., 1983, for a summary) and boys, relative to girls, tend to be more facile with objects, as indicated by their performance on the block design portion of the Wechsler Preschool and Primary Scale of Intelligence (WPPSI) (Caldera et al., 1999).

Considering the ways in which objects are used in construction, exploration, play, and tool use should clarify this confusion by presenting descriptive information on boys' and girls' object use. Specifically, we posit that girls will spend more time in construction than boys. Boys, on the other hand, will spend more time playing with objects and using objects as tools, especially in tool use that reflects male stereotypes (use of objects as weapons and for banging). This hypothesis is examined in an empirical study reported next.

We also examine the relation between object use, especially during play, and problem solving. We present an empirical study relating boys' and girls' observed object use during their classroom free play and to their subsequent ability to use tools to solve problems. Specifically, we examine children's ability to use tools to solve "convergent problems" and "divergent problems." Both types of problems have been examined extensively in the child development literature. We examine two types of convergent problems and one divergent task. More detail on method is provided below; here we give a general overview of the tasks.

In the first convergent task, children are presented with an assembled tool and two components of tools and asked to use the objects to retrieve a lure. In the second convergent task, they are presented with components of a tool that must be assembled in order to retrieve the lure. In the divergent task, children are given a conventional object and asked to generate novel uses of it (i.e., associative fluency).

Research in this area has a mixed history. The classic study of Sylva and colleagues (1976) found that play did affect performance on a tool construction and use task, but other researchers have not replicated these results when double-blind procedures were used (Simon & Smith, 1983;

Vandenberg, 1980). Similarly, the history of research on relations between play and associative fluency has been mixed. In two frequently cited papers, Dansky and Silverman (1973, 1975) reported that play facilitated associative fluency in preschoolers. However, these results did not replicate under double-blind conditions (Smith & Whitney, 1987).

In the study reported next, we present data on relations between different types of object use and children's ability to use objects to solve different types of problems.

METHOD

Participants and Procedures

The 35 children (17 females, 18 males), 3–5 years of age, involved in this study attended the Shirley Moore Laboratory Nursery school located on the University of Minnesota campus. A subsample was involved in a set of procedures that examined children's facility with using and making tools to solve problems. This group consisted of 20 children (10 females, 10 males) from two classrooms.

Procedures involved direct observation of two retrieval tasks (one involving a connected tool, a second involving the construction of a tool from unconnected components); an associative fluency task; and a spatial IQ task (WPPSI block design).

Observations were conducted three afternoons per week for one entire school year. Classrooms included a variety of activity centers and play areas that were equipped with a wide assortment of play materials such as blocks, dress-up clothes, stuffed animals, dolls, trucks, musical instruments, puzzles, board games, clay, art materials (for painting, writing, drawing, and cutting), water tables, sand tables, book areas, computers (installed with age-relevant software programs), and printers. The playground area included a large sand area, a large climbing apparatus, and swings, as well as an assortment of tricycles, toy wagons, and shovels. Classrooms were equipped with observation booths, from which all indoor observations were conducted. Outdoor observations were conducted while standing on the side of the playground area.

Observations consisted of 1-minute focal child sampling with continuous recording (Pellegrini, 2004) during periods of free play. During each 1-minute interval, all relevant behaviors of the focal child were continuously recorded, in the order in which they occurred, on a coding sheet. Here we report four dimensions of the focal child's behavior with objects: exploration, play, construction, and tool use. Behaviors that did not involve objects were coded as "other." Tool use was further coded according to the following nine subcategories: reach, weapon, body aid, pour/wipe, bang, empty/fill/dig/poke, art/write, transport/carry, and throw/kick. Descriptions of each of the four categories and nine subcategories are provided in Table 6.1.

Children were asked to perform the tasks individually in the hall outside their classrooms. Their performance was videotaped (first two tasks) or audiotaped (third task).

The connected tool retrieval task involved selecting a tool with which to retrieve a toy dinosaur that had been placed out of reach of the student. The materials for the first task consisted of three potential retrieval tools: a plastic toy hoe (25 inches long and 5 inches wide), a plastic rake head without a handle (10½ inches long and 6 inches wide), and a plastic toy rake handle without the rake head (17½ inches long). A plastic toy dinosaur (7 inches long and 3 inches high) was used as the item to be retrieved.

TABLE 6.1. Object Use Categories and Tool Use Subcategories

Object use	Explanation and example
Exploration	"What can it do?" Flat or negative affect. Sniff; squeeze; shake; rotate, bring to eyes; drop.
Play	"What can I do with it?" Positive affect. Pretending (e.g., cues in language or gestures).
Construction	Building something that is end oriented (not play).
Tool use	Using an object as a means to an end; to get something done.
Reach	Extend reach; rake; insert and probe; stack and climb.
Weapon	Throw; brandish; wave; threaten with.
Body aid	Support self (e.g., using an object as a crutch).
Pour/wipe	Control liquids (e.g., using an object to pour liquid).
Bang	Using an object to make a noise; crashing or slamming an object for the sake of making noise.
Empty/fill/dig/poke	Using an object to fill or empty containers; using an object to jab or prod.
Art/write	Using an object to write, draw, paint, etc.
Transport/carry	Using an object to transport or carry something; or transporting, carrying, or holding an object.
Throw/kick	Throwing or kicking an object (but not as a weapon).

In the connected task children were told:

> "In this game I want you to figure out a way to use these things, or one of them, to get the dinosaur at the other end of the table. You can use these things in any way to get the dinosaur. You cannot get out of your seat, however. Take your time and try. Go ahead and let's play the game."

Children were given a series of hints if they could not perform the task:

> "Why don't you try reaching with one of these things? [Next they were told:] Why don't you try this? [The experimenter pointed at the connected rake.]"

For the unconnected task, children were told:

> "In this game I want you to figure out a way to use these things to get the dinosaur. You can use these things in any way you want to get the dinosaur. You cannot get out of your seat, however. Go ahead, take your time and try."

Hints were provided, as needed:

> "Can you think of a way to use *some* of these things? Can you try using the *round yellow pieces and the sticks* to help you? Can you try using the *round yellow pieces and the sticks together* to help you? Can you put one stick into each end of the yellow round piece to make a *longer* stick? I will hold a stick. Can you put the round yellow piece on the end of it? Can you use the other sticks and round yellow pieces to make it *longer?*"

The following dimensions of each child's performance were scored: the total time (in seconds) needed by the child to use one or more of the objects to successfully retrieve the dinosaur; the number of hints provided to the child while completing the task; and the number of swipes (e.g., attempts to use one or more of the objects to retrieve the dinosaur).

The unconnected tool retrieval task involved constructing a tool with which to retrieve the toy dinosaur that again had been placed out of reach of the student. The construction materials for the second task consisted of selected pieces from a Tinker Toy set: two plastic Tinker Toy sticks (10½ inches long and ⅜ inch in diameter); two round plastic Tinker Toy pieces (2 inches in diameter and ½ inch wide) with a hole in each face, into which a stick could be fitted and masking tape used to cover the remaining holes around the edges; and an apparatus constructed of several Tinker Toy pieces. The constructed apparatus consisted of a plastic stick (10½ inches long) connected perpendicularly to a 5-inch plastic stick, with a round plastic piece (2 inches in diameter and ½ inch wide) fitted into each end of the 5-inch stick. Again, a plastic toy dinosaur (7 inches long and 3 inches high) was used as the item to be retrieved.

After completion of the first task, the retrieval dinosaur was again placed on the table. The Tinker Toy pieces for the second task were placed on the table in front of the child in the following order (from left to right): the constructed apparatus; a 10½-inch stick; a round piece; a 10½-inch stick; and a round piece.

The child was told:

> "In this next game, I want you to figure out a way to use these things to get the dinosaur. You can use these things in any way you want. However, you cannot get out of your seat. Go ahead—take your time and try."

If the child did not make any attempt to interact with the objects (after 1–1½ minutes had elapsed), or if the child appeared confused, the experimenter provided the child with a hint.

The hints were provided in a sequential order but in ways that were appropriate to the phase of the task in which the child was engaged. The hints also provided the child with gradually more specific help in accomplishing the second task. The hints provided for this tool construction/retrieval task, consisted of the following:

1. "Can you think of a way to use some of these things to get the dinosaur?"
2. "Can you try using a round piece to help you?"
3. "Can you try using the round pieces and the sticks together to help you?"
4. "Can you put one stick into each end of the round piece to make a longer stick?"
5. "I will hold this stick. [The experimenter picks up a stick for the child.] Can you put the round piece on the end of it?"
6. "Can you use the other sticks and round pieces to make it longer?"
7. The experimenter connected the pieces for the child and had him or her use the experimenter-constructed pieces to retrieve the dinosaur.

At the end of the second task, the retrieval dinosaur and the Tinker Toy pieces were removed from the table and the video camera turned off.

The following dimensions of each child's performance were scored: the total time (in seconds) needed by the child to construct a tool and use it to successfully retrieve the dinosaur; the number of hints provided to the child during completion of the task; and the number of swipes (e.g., attempts to use the constructed tool and/or one or more of the Tinker Toy pieces to retrieve the dinosaur).

The third task, measuring associative fluency, involved asking children to generate novel uses for one of three common household objects (Pellegrini & Greene, 1980; Wallach, 1970). One of the following objects was placed on the table in front of the child: a paper cup, a plastic spoon, or a marker (in counterbalanced order). Responses were used to create two measures: conventional uses and novel uses of everyday objects. We analyzed only the novel uses, defined as those that reflected an unusual or atypical way of using the targeted object.

After each child had completed these three tasks, the WPPSI Block Design Test was administered. Following the directions of the subtest, each child was asked to use a set of colored blocks to reproduce up to 14 different block patterns, presented to him or her one at a time by the examiner. Two attempts were allowed to reproduce each design.

Scores were based on the number of block designs reproduced correctly, the number of block designs reproduced correctly on the first attempt, and (on selected designs) the number of block design patterns completed correctly within a specified time limit. At the completion of the test, each child's Block Design raw score was converted to a Block Design scaled score, using a table provided in the WPPSI manual.

RESULTS/DISCUSSION

First we present descriptive information on the occurrence of different forms of object use across the school year; then we analyze children's use of tools to solve problems.

Observations of Object Use

Descriptive statistics are displayed in Table 6.2. The numbers in this table reflect the rate of occurrence of each category for each focal child observation. As can be seen, the rates of occurrence for play and tool use were similar, approximately one per observation. Total tool use and play accounted

TABLE 6.2. Descriptive Statistics for Object Use by Sex

	Males (n = 18)		Females (n = 17)			
Category	M	SD	M	SD	t	p
Explore	0.11	.15	0.005	.007	−1.40	.16
Play	1.40	.52	0.96	.50	−2.55	.01
Construction	0.50	.40	0.98	.40	2.91	.006
Other behavior	1.68	.67	1.66	.47	.083	.93
Total tool	1.21	.62	0.97	.41	−1.32	.19
Reach	0.003	.10	0.003	.01	−1.35	.18
Weapon	0.01	.002	0.00	.00	−1.73	.09
Body aid	0.0002	.001	0.00	.00	−.97	.33
Pour	0.19	.21	.009	.14	−1.55	.13
Bang	0.009	.12	.001	.003	−2.57	.01
Empty	0.11	.23	.006	.008	−.82	.41
Art	0.23	.23	0.36	.28	1.4	.15
Transport	0.47	.32	0.39	.26	−.77	.44
Throw	0.004	.007	0.003	.006	−.52	.60

for 24% and 26%, respectively, of all observed behavior, with males engaging in each form of behavior more than females. From these descriptive data we conclude that tool use, like play, occurs at a moderate level during children's play time in nursery school.

In comparing time budgets spent in tool use between human juveniles and chimpanzees (McGrew, 1981), we find that humans spend substantially more (c. 25% vs. 10–15%). This difference probably reflects the enriched nursery school environment in which these children were observed. The classrooms and the playground were equipped with various and numerous objects and tools with which children were encouraged to play. A more valid comparison would entail observing children across their whole day, in and out of school. When such a comparison is made (Bock, 1999), the time budgets are more in line with the chimpanzee case (c. 16–17% at peak).

Sex Differences in Occurrence of Object Use

Sex differences for exploration, play, construction, and tool use (individual and aggregate categories) were analyzed with t-tests for independent samples. Significant sex differences were observed for construction (females more), play, and the bang tool subcategory use (males more), but not for the aggregate tool category. The sex difference for tool as weapon subcategory was only marginally significant.

The sex difference favoring females for construction is consistent with extant research (Rubin et al., 1983). Girls' uses of objects to build things often took the form of making puzzles and doing art activities, all relatively sedentary activities. Boys, on the other hand, used objects in play and marginally more as tools. Boys' play with objects was often embedded in the context of fantasy play, in which objects were used to enact superhero themes, consistent with the early finding of Saltz, Dixon, and Johnson (1977). Also consistent with the comparative literature, we found that males, more than females, tended to use objects as weapons.

Exploration, Play, Construction, and Tool Use as Distinct Categories

Next, we examined the co-occurrence of exploration, construction, object play, and tool use among preschool children. The correlation coefficients among age, play, exploration, and different forms of tool use are displayed in Table 6.3.

First, these analyses reinforce the position that exploration, construction, play, and tool use are separate constructs. Specifically, exploration and play were positively but not significantly correlated. Furthermore, play and construction were negatively and significantly correlated. Regarding play and tool use, play and weapon use were positively and significantly correlated, whereas play and art were negatively and significantly correlated; play was negatively, and not significantly, correlated with total tool use.

These results reinforce the importance of differentiating children's uses of objects. Consistent with Piaget's formulations, construction and play were negatively intercorrelated; consistent with Hutt, exploration and play were independent. Total tool use was not related to play, exploration, or construction.

Using Objects as Tools to Solve Problems

The descriptive statistics for all tasks are displayed in Table 6.4, and the intercorrelations among these measures are presented in Table 6.5.

No significant sex differences were observed on any of the three tasks, though the directions favored the females in the connected and associative fluency tasks. The lack of sex differences for convergent (both the tool

TABLE 6.3. Intercorrelation between Measures of Object Use

	2	3	4	5	6	7	8	9	10	11	12	13	14
Age 1	0.13	0.05	**–0.35**	–0.06	–0.28	0.05	–0.17	0.05	0.06	–0.31	**–0.39**	–0.24	0.01
Explore 2		–0.007	–0.1	0.04	–0.12	0.19	0.01	–0.15	–0.30	–0.23	0.23	–0.20	–0.04
Play 3			0.15	**0.39**	0.13	–0.08	0.13	0.09	–0.26	–0.01	–0.02	–0.07	**–0.36**
Reach 4				–0.08	**0.93**	0.04	0.20	0.05	0.09	0.08	0.08	**0.32**	–0.13
Weapon 5					–0.05	–0.08	**0.57**	–0.14	–0.01	0.09	0.12	0.10	–0.29
Body 6						0.09	0.25	0.06	0.16	–0.02	0.13	**0.30**	–0.09
Pour 7							**0.26**	**0.46**	**0.33**	–0.11	–0.26	**0.62**	–0.22
Bang 8								–0.15	0.26	0.03	0.09	**0.44**	**–0.37**
Empty 9									0.004	0.04	–0.10	**0.48**	–0.03
Art 10										–0.18	–0.28	**0.53**	0.10
Transport 11											–0.12	**0.43**	–0.07
Throw 12												–0.16	–0.05
Total Tool 13													–0.18
Construct 14													

Note. Values underlined: $p < .10$; **values in bold: $p < .05$**; **values in bold and underlined: $p < .01$**

TABLE 6.4. Descriptive Statistics for Objects Used to Solve Problems by Sex

	Males (n = 10)		Females (n = 10)		t	
Category	*M*	*SD*	*M*	*SD*	(df = 18)	*p*
Associative fluency	3.10	2.38	5.00	4.42	1.19	.24
Connected retrieval task						
Time (in sec)	33.70	35.48	27.50	22.09	−0.46	.64
Hints	1.50	.85	1.30	.48	−0.64	.52
Swipes	1.70	1.57	1.70	.82	0	1.00
Unconnected retrieval task						
Time (in sec)	102.00	73.20	138.20	114.91	0.84	.41
Hints	2.20	1.23	3.80	2.90	1.60	.12
Swipes	3.10	1.37	4.60	3.44	1.28	.21

tasks) and divergent (associative fluency) problem-solving tasks are consistent with earlier work (Pepler & Ross, 1981).

Measures of performance within each of the two retrieval tasks were highly interrelated, but measures of performance between tasks were not interrelated; thus, performance on each task was specific to that task. With this said, it would have been more informative to examine correlations between object use and problem solving within each sex, had the sample size

TABLE 6.5. Intercorrelations between Uses of Objects to Solve Problems

	2	3	4	5	6	7	8
Age 1	0.27	0.17	0.20	0.03	−0.10	−0.10	−0.07
Associative fluency 2		0.08	0.07	0.36	−0.41	−0.13	0.02
Connected tool task time 3			**0.89**	**0.64**	0.20	0.001	0.30
Connected tool task hint 4				**0.47**	0.15	−0.10	0.12
Connected tool task swipe 5					0.008	0.24	**0.47**
Unconnected tool task time 6						**0.74**	**0.58**
Unconnected tool task hint 7							**0.82**
Unconnected tool task swipe 8							

Note. n = 20. Values underlined: $p < .10$; **values in bold: $p < .05$**; **values in bold and underlined $p < .01$**.

permitted. That females, more than males, engaged in construction and that males, more than females, engaged in play and tool use suggest that there may have been different predictive relations for each sex.

Next we examined the extent to which observed object use predicted performance on the objects as tools tasks, when spatial intelligence was controlled. These analyses should provide some insight into the degree to which time spent in different sorts of object use relates to solving problems with objects, independent of more general spatial intelligence. Table 6.6 displays these partial correlations.

Observed exploration, construction, and tool use were negatively related to measures of time, number of hints needed, and number of swipes at the lure, indicating that these measures of observed object use predicted facility in the lure retrieval tasks. Specific to construction, more time spent in construction related to less time and fewer hints needed to solve the connected tool problem. It is probably the case that children are using similar

TABLE 6.6. Partial Correlations between Observed Object Use and Performance on Using Objects as Tools, Controlling for Spatial Intelligence

	Performance						
Object use	Associative fluency	Connected tool task time	Connected tool task hint	Connected tool task swipe	Unconnected tool task time	Unconnected tool task hint	Unconnected tool task time
Explore	0.04	−0.21	−0.21	−0.24	−0.16	−0.38	−0.35
Play	−0.18	0.35	0.33	0.33	0.10	−0.15	−0.11
Reach	−0.31	−0.20	−0.15	−0.22	−0.26	−0.22	−0.28
Weapon	0.07	**0.70**	**0.64**	**0.67**	0.09	−0.14	0.002
Body	−0.26	−0.19	−0.15	−0.17	−0.22	−0.19	−0.25
Fluid	−0.33	−0.26	−0.12	−0.32	−0.004	−0.24	−0.14
Bang	−0.12	0.42	**0.54**	0.06	0.003	−0.43	−0.22
Empty	−0.23	−0.44	**−0.47**	−0.16	−0.24	−0.11	−0.02
Art	−0.21	−0.07	0.01	−0.18	−0.15	−0.19	−0.02
Transport	0.01	0.21	0.26	0.17	−0.16	−0.07	0.01
Throw	0.12	−0.07	−0.24	−0.04	−0.32	−0.33	−0.30
Total tool	−0.34	−0.01	0.12	−0.14	−0.32	−0.45	−0.25
Construct	**0.52**	**−0.45**	**−0.46**	−0.17	−0.30	−0.06	−0.18

Note. $n = 20$. $df = 16$. Values underlined: $p < .10$; **values in bold: $p < .05$; values in bold and underlined: $p < .01$.**

skills in both construction and problem solving and that spending time in construction provides important practice and learning opportunities for problem solving. Both tasks involve using objects in ends-oriented ways of solving problems. Total tool use was also a significant predictor of problem solving, but in this case, solving the unconnected tool task. More time spent in total tool use related to fewer hints needed to solve the problem.

The different patterns for connected and unconnected tasks merit further discussion. First, as displayed in Table 6.5, performance on one task was not related to performance on the other, though measures of performance within each task were interrelated. Second, the connected task was less demanding than the unconnected task, as evidenced by the difference in time needed to solve each problem. Third, the magnitudes of the correlations between observed tool use and performance on the unconnected task were higher than on the connected task. The pattern for observed construction was the opposite: Correlations were higher (in two of the three cases) between time in construction and performance on the connected compared to the unconnected tasks. With regard to the unconnected tool task, it is probably the case that the specific skills needed to solve the problem took more time and practice than those needed in the easier connected task.

Interestingly, the more time spent in two subcategories of tool use, banging and weapon, related to more time and hints needed to solve the connected problem. Indeed, the magnitudes of the correlations for the weapon subcategory were especially high. The weapon subcategory was also significantly related to the play category, suggesting that children who used weapons as tools also engaged in fantasy. In our observations, there were no cases of girls using tools as weapons; thus, this was a male phenomenon.

The direction of the correlations for this subcategory and play on performance of the connected task suggests that fantasy does not predict performance on this sort of task. It would be interesting, however, to examine the extent to which the weapon subcategory related to a more closely related task, such as throwing an object at a target. We would predict that males' use of tools as weapons does relate to this sort of tool-use task. Consistent with this hypothesis, sex differences on children's performance of target tasks (favoring boys as well as those girls with high levels of androgen, relative to control girls) but not on mental rotation tasks have been reported (Hines et al., 2003).

Why did time spent in construction predict both performance on the divergent problem and the connected convergent problem? Construction is an ends-oriented behavioral category. In the process of constructing, children typically coordinate means toward some end product. This set of skills was sufficient to solve the simpler connected retrieval task and the associative fluency task but less effective with the more complex unconnected retrieval task. A more diverse and specific set of skills, reflected in the total tool category, was necessary to solve the unconnected task.

CONCLUSION

In this chapter we have reviewed the literature on children's object use and presented data on children's uses of objects in exploration, construction, tool use, and play with objects. Results were unequivocal in showing that these categories are distinct and should be treated as such. Correspondingly, these categories have separate developmental trajectories and consequences. Much more work is needed, however, to understand the ways in which individuals develop, particularly observational work documenting the time spent in different sorts of object use. Observations outside of the school setting are especially needed; we need to study children in those contexts in which they are developing (McCall, 1977; Wright, 1960). Schools are one important developmental context, but they represent only a limited portion of children's days. We need observations in a variety of setting to make more accurate estimates about time budgets of object use.

In schools, even in those where children are permitted and encouraged to engage in free play and relatively unfettered peer interaction, policy often limits certain behaviors. For example, many schools have prohibitions against certain kinds of object use (for example, pretending an object is a gun or another sort of weapon) and play (for example, rough-and-tumble play). On the other hand, in some schools, as in the present study, the object-rich environments and curriculum likely result in overestimates of object use, relative to out-of-school object use. Both of these sorts of constraints restrict the samples of behavior available to the researcher.

Examining sex differences in object and tool use remains a complex venture. Evidence from research on chimpanzees certainly highlights this complexity but also provides some guidance. Female chimpanzees and human juveniles alike are facile at using objects for constructive purposes. In the former case, this is often done in the context of gathering. Despite this bias, females did not outperform males on either of our convergent tool tasks, both of which involved constructing tools to gather an object. As noted above, the direction of the means in two of the three tasks favored females, though the differences were not statistically significant. Future research should test for sex differences on tasks using tools to gather and process resources in more varied ways.

The anthropological work of Bock (Chapter 10, this volume) provides some guidance here. Juvenile females' tool use, for example, was assessed in the processing of grain—a task they were expected to perform as adults. Perhaps a female-oriented task could involve using a stick to locate a stimulus (such as a fruit or vegetable) embedded in a complex field of other stimuli. Similarly, a male-oriented task might test ability to saw a piece of wood accurately or to throw a stick at a target.

ACKNOWLEDGMENTS

We acknowledge the support and cooperation of Lynn Galle and the teachers, staff, children, and families of the Shirley Moore Laboratory School at the University of Minnesota, and the comments of P. K. Smith.

REFERENCES

Archer, J. (1992). *Ethology and human development.* Hemel Hempstead, UK: Harvester Wheatsheaf.

Bates, E., Carlson-Luden, V., & Bretherton, I. (1980). Perceptual aspects of tool use in infancy. *Infant Behavior and Development, 3*, 127–140.

Belsky, J., & Most, R. (1981). From exploration to play: A cross-sectional study of infant free-play behavior. *Developmental Psychology, 17*, 630–639.

Blurton Jones, N. G., Hawkes, K., & O'Connell, J. F. (1997). Why do Hadza children forage? In N. L. Segal, G. E. Weisfeld, & C. C. Weisfeld (Eds.), *Uniting psychology and biology: Integrative perspectives on human development* (pp. 279–313). Washington, DC: American Psychological Association.

Bock, J. (1999, April). *The socioecology of children's activities: A new model for work and play.* Paper presented at the biennial meetings of the Society for Research in Child Development, Albuquerque, NM.

Boyd, R., & Richerson, P. J. (1985). *Culture and the evolutionary process.* Chicago: University of Chicago Press.

Caldera, Y. M., Culp, A. M., O'Brien, M., Truglio, R. T., Alvarez, M., & Huston, C. (1999). Children's play preferences, constructive play with blocks, and visual–spatial skills: Are they related? *International Journal of Behavioral Development, 23*, 855–872.

Caro, T. (1988). Adaptive significance of play: Are we getting closer? *Trends in Ecology and Evolution, 3,* 50–54.

Chen, Z., & Siegler, R. S. (2000). Across the great divide: Bridging the gap between understanding of toddlers' and older children's thinking. *Monographs of the Society for Research in Child Development, 65* (Serial No. 261).

Connolly, K., & Dalgleish, M. (1989). The emergence of a tool-using skill in infancy. *Developmental Psychology, 23*, 894–912.

Connolly, K., & Elliott, J. (1972). The evolution and ontogeny of hand functions. In N. Blurton Jones (Ed.), *Ethological studies of child behaviour* (pp. 329–384). Cambridge, UK: Cambridge University Press.

Dansky, J., & Silverman, I. W. (1973). Effects of play on associative fluency in preschool-age children. *Developmental Psychology, 9*, 38–43.

Dansky, J., & Silverman, I. W. (1975). Play: A general facilitator of associative fluency. *Developmental Psychology, 11*, 104.

Erickson, E. H. (1977). *Toys and reason.* New York: Norton.

Fagen, R. (1981). *Animal play behavior.* New York: Oxford University Press.

Fein, G. (1981). Pretend play in childhood: An integrative review. *Child Development, 52*, 1095–1118.

Henderson, B. B., & Moore, S. G. (1979). Measuring exploratory behavior in young children: A factor analytical study. *Developmental Psychology, 15*, 113–119.

Hines, M., Fane, B. A., Pasterski, V. L., Mathews, G. A., Conway, G. S., & Brooks, C. (2003). Spatial abilities following prenatal androgen abnormality: Targeting and mental rotations performance in individuals with congenital adrenal hyperplasia. *Psychoneuroendocrinology*, 28, 1010–1026.

Hutt, C. (1966). Exploration and play in children. *Symposia of the Zoological Society of London, 18*, 61–81.

Johnson, J. E., & Ershler, J. (1981). Developmental trends in preschool play as a function of classroom program and child gender. *Child Development, 52*, 995–1004.

Johnson, J. E., Ershler, J., & Bell, C. (1980). Play behavior in a discovery-based and a formal education preschool program. *Child Development, 51*, 271–274.

Liddell, C., Kvalsvig, J., Strydom, N., Qotyana, P., & Shabalala, A. (1993). An observational study of 5-year-old South African children in the year before school. *International Journal of Behavioral Development, 16*, 537–561.

Martin, P. (1982). The energy cost of play: Definition and estimation. *Animal Behaviour, 30*, 294–295.

Martin, P., & Caro, T. (1985). On the function of play and its role in behavioral development. In J. Rosenblatt, C. Beer, M.-C. Bushnel, & P. Slater (Eds.), *Advances in the study of behavior* (Vol. 15, pp. 59–103). New York: Academic Press.

McCall, R. B. (1977). Challenges to a science of developmental psychology. *Child Development, 48*, 333–344.

McGrew, W. C. (1981). The female chimpanzee as a female evolutionary prototype. In F. Dahlberg (Ed.), *Woman the gatherer* (pp. 35–73). New Haven, CT: Yale University Press.

McGrew, W. C. (1992). *Chimpanzee material culture.* Cambridge, UK: Cambridge University Press.

Montessori, M. (1964). *The Montessori method.* New York: Schocken Books.

Pellegrini, A. D. (2004). *Observing children in their natural worlds: A methodological primer* (2nd ed.). Mahwah, NJ: Erlbaum.

Pellegrini, A. D., & Greene, H. (1980). The use of a sequenced questioning paradigm to facilitate associative fluency in preschoolers. *Journal of Applied Developmental Psychology, 1*, 189–200.

Pellegrini, A.D., Horvat, M., & Huberty, P.D. (1998). The relative cost of children's physical activity play. *Animal Behaviour, 55*, 1053–1106.

Pellegrini, A. D., & Perlmutter, J. (1989). Classroom contextual effects on children's play. *Developmental Psychology, 25*, 289–296.

Pepler, D., & Ross, H. (1981). The effects of play on convergent and divergent problem solving. *Child Development, 52*, 1202–1210.

Piaget, J. (1962). *Play, dreams, and imitation in childhood* (C. Gattengno & F. M. Hodgson, Trans.). New York: Norton. (Original work published 1951)

Piaget, J. (1983). Piaget's theory. In W. Kessen (Ed.), *Handbook of child psychology: History, theory, and methods* (pp. 103–128). New York: Wiley.

Power, T. G. (2000). *Play and exploration in children and animals.* Mahwah, NJ: Erlbaum.

Rubin, K. H., Fein, G., & Vandenberg, B. (1983). Play. In E. M. Hetherington (Ed.), *Handbook of child psychology: Vol. IV. Socialization, personality and social development,* (pp. 693–774). New York: Wiley.

Rubin, K. H., Maioni, T., & Hornung, M. (1976). Free play in middle and lower class preschoolers: Parten and Piaget revisited. *Child Development, 47*, 414–419.

Rubin, K.H., Watson, R., & Jambor, T. (1978). Free-play behaviors in preschool and kindergarten children. *Child Development, 49*, 534–546.

Saltz, E., Dixon, D., & Johnson, J. (1977). Training disadvantaged preschoolers on various fantasy activities: Effects on cognitive functioning and impulse control. *Child Development, 48*, 367–380.

Simon, T., & Smith, P. K. (1983). The study of play and problem solving in preschool children. *British Journal of Developmental Psychology, 1*, 289–297.

Smilansky, S. (1968). *The effects of sociodramatic play on disadvantaged preschool children.* New York: Wiley.

Smith, P. K., & Simon, T. (1984). The study of play and problem-solving in preschool children: Methodological problems and new directions. In P. K. Smith (Ed.), *Play in animals and humans* (pp. 199–216). Oxford, UK: Blackwell.

Smith, P. K., Takhvar, M., Gore, N., & Vollstedt, R. (1986). Play in young children: Problems of definition, categorization, and measurement. In P. K. Smith (Ed.), *Children's play* (pp. 39–55). London: Gordon and Breach.

Smith, P. K., & Whitney, S. (1987). Play and associative fluency: Experimenter effects may be responsible for previous positive findings. *Developmental Psychology, 23*, 49–53.

Smitsman, A. W., & Bongers, R. M. (2003). Tool use and tool making: A developmental action perspective. In J. Valsiner & K. J. Connolly (Eds.), *Handbook of developmental psychology* (pp. 172–193). London: Sage.

Sutton-Smith, B., Rosenberg, B. G., & Morgan, E. F., Jr. (1963). Development of sex differences in play choices during adolescence. *Child Development, 34*, 119–126.

Sylva, K., Bruner, J., & Genova, P. (1976). The role of play in the problem-solving of children 3–5 years old. In J. Bruner, A. Jolly, & K. Sylva (Eds.), *Play: Its role in development and evolution* (pp. 244–261). New York: Basic Books.

Tanner, N. M. (1987). The chimpanzee model revisited and the gathering hypothesis. In W. G. Kinzey (Ed.), *The evolution of human behavior: Primate models* (pp. 3–27). Albany: State University of New York Press.

Tinbergen, N. (1963). [On the aims and methods of ethology]. *Zeitschrift für Tierpsychologie, 20,* 410–413.

Tomasello, M., & Call, J. (1997). *Primate cognition.* New York: Oxford University Press.

Vandenberg, B. (1980). Play, problem solving, and creativity. In K. Rubin (Ed.), *Children's play* (pp. 49–68). San Francisco: Jossey-Bass.

van Leeuwen, L., Smitsman, A., & van Leeuwen, C. (1994). Affordances, perceptual complexity, and the development of tool use. *Journal of Experimental Psychology: Human Perception and Performance, 20,* 174–191.

Wallach, M. (1970). Creativity. In P. H. Mussen (Ed.), *Carmichael's manual of child psychology* (Vol. 1, 3rd ed., pp. 1211–1272). New York: Wiley.

Wrangham, R. (1987). The significance of African apes for reconstructing human social evolution. In W. G. Kinzey (Ed.), *The evolution of human behavior: Primate models* (pp. 51–71). Albany: State University of New York Press.

Wright, H. (1960). Observational child study. In P. H. Mussen (Ed.), *Handbook of research methods in child development* (pp. 71–139). New York: Wiley.

PART IV

FANTASY

CHAPTER 7

Fantasy Play in Apes

JUAN-CARLOS GÓMEZ AND BEATRIZ MARTÍN-ANDRADE

In this chapter we review the existing evidence about the occurrence of fantasy play in nonhuman primates, especially anthropoid apes. We begin with a discussion of how imaginative or fantasy play has been defined by different authors in an attempt to derive objective criteria for the identification of this kind of behavior and the evaluation of claims by students of primates. We then review the purported observations of behaviors that have been claimed to qualify as fantasy play by different authors studying nonhuman primates; we organize our review into observations of natural behaviors, observations in captive settings, and observations in hand-reared primates and "linguistic" apes (i.e., apes that have been trained in the use of artificial symbols). We conclude that the most convincing (or at least the more human-like) examples of fantasy play are enacted by symbolically trained apes, although the behaviors are not necessarily the product of direct training, but rather a by-product of the training process. On the other hand, it is possible to observe some prerequisites or precursors of imaginative play in some behavior of untrained apes. Based upon this evidence, we propose that human fantasy play may have evolved from a combination of precursor behaviors and abilities related to symbolic and explicit representation, only some of which are shown by modern apes.

Symbolic play is one of the trademarks of human childhood. From as early as the end of the first year of life, human babies engage in actions that resemble behaviors performed by adults or by themselves in "serious" contexts but that do not appear to be intended as serious attempts at performing the real actions. For example, a baby may take a spoon, put it in an empty cup, and then take the spoon to his or her own mouth or someone else's mouth (see Smith, Chapter 8, this volume). Or a baby may lie on a cushion and close his or her eyes, smiling as if asleep. During the second and third years of life, fantasy play blossoms. Children feed, clean, scold, put to sleep, and read tales to their dolls, or pretend that they are horses or cats, to cite but a few examples every parent can constantly observe in their children.

However, in young monkeys and apes symbolic play appears to be an extremely rare occurrence, if it occurs at all. As Fagen (1981, p. 123) put it in his encyclopedic review of animal play some years ago, young chimpanzees and apes could "pretend to be leopards, baboons or field primatologists," but they have never been described engaging in such patterns of behavior, "suggesting that this level of symbolic complexity most likely lies beyond the reach of any species on this planet but our own." In this chapter we discuss to what extent evidence gathered in recent years questions this conclusion reached by Fagen.

DEFINING FANTASY AND IMAGINATIVE PLAY

Fantasy or pretend play is difficult to define precisely. This difficulty derives, on the one hand, from the difficulty in defining play in general (Fagen, 1981) and, on the other, from a specific difficulty in defining *pretending* and *imagination*. This type of play consists of "acting as if" something were the case, when it actually is not, and doing so outside a directly functional context. This definition excludes from our review instances of deception, whereby a person or animal pretends with the aim of effectively misleading others and obtaining a benefit as a result; for example, a subordinate ape refraining from looking at a piece of food while a dominant is present, might be pretending that there is no food (see Byrne & Whiten, 1988; Mitchell, 2002a, for reviews and discussions of deceptive pretense), but is not engaging in pretend or fantasy *play*.

We use the labels "symbolic," "fantasy," "imaginative," and "pretend" play as essentially synonymous, although some finer distinctions could be made. For example, fantasy might be defined differently from pretense as the ability to use symbols to invent things that do not exist. In this sense, fantasy play might involve more than symbolic or pretend play, in that acting out pretenses of things that exist in the real world (e.g., food, a car) would not qualify as fantasy, whereas acting out things that do not exist, such as a monster or an invented person, would. In this chapter we use the wider meaning in which fantasy and symbolic or pretend play broadly coincide.

Pretend or "as if" actions usually reproduce only some components of their serious counterparts. For example, a child pretending to feed a doll may use something different from a spoon and no food at all, and his or her actions may not reproduce exactly those used in real feeding, but occur in a condensed or exaggerated way. A typical way of pretending consists of using an object as if it were a different one; for example, using a stick as the "spoon" to feed a doll. This is why this form of play is called "symbolic": It consists of actions carried out with the intention that they *stand for* other actions, objects, or situations that are not present in that moment.

The ultimate criterion for identifying an action as an instance of imaginative play is precisely the existence of such an online symbolic intent superimposed upon the literal action. Since this intention to pretend cannot be directly observed but must be inferred from behavioral signs such as smiles or the structure of actions, symbolic play may be difficult and contentious to identify, especially in nonverbal organisms, such as animals or very young children, who cannot linguistically declare their playful intentions.

Even with human children, especially in their earliest stages of play development, it is frequently difficult to be sure that a particular action is intended as pretense (Huttenlocher & Higgins, 1978). To overcome this problem we can use the structure of the playful action and the context in which it is produced as criteria to identify fantasy play. Pretend actions are usually subject to some sort of schematization that renders them structurally different from their serious counterparts. In the human literature, authors such as Dunn and Wooding (1977) or Fein and Apfel (1979) suggest the following criteria to identify "nonliteral" actions: Familiar activities are carried out without the necessary objects or materials or outside their usual context (e.g., feeding without actual food or feeding implements); actions may not be completed until attaining their usual results (e.g., a cup with actual water is held close to the lips while making drinking sounds, but the water is not consumed); inanimate objects are treated as animate; one object is used instead of another; the child "impersonates" another person or an object (e.g., acting as if he or she were a firefighter or a train).

There is no single set of criteria to identify symbolic or imaginative play, or indeed most other forms of play. Most sets of criteria are similar to the above (see Mitchell, 2002b), although there is space for variation and even some contradictions or paradoxes. For example, although decontextualization (e.g., cleaning when there is no dirt) is a usual feature of symbolic play, sometimes it is its opposite that causes the impression of pretend play: For example, a toddler who takes her doll to her nappy changing mat before engaging in some clumsy cleaning of the doll's bottom with a piece of paper causes a stronger impression of engaging in imaginative play because she is providing additional appropriate context (our personal observations).

This state of affairs makes the task of looking for instances of imaginative play in nonhuman primates all the more difficult. Indeed it also makes difficult the identification of the precise moment in which genuine imaginative play appears in human babies and its cognitive mechanisms.

ONTOGENY AND NATURE OF IMAGINATIVE PLAY

Different accounts of the course of fantasy play development emphasize different milestones and features (Power, 2000). Piaget (1945) provided an influential model in which genuine pretend play was at work by the end of the second year of life, but it was preceded by many ontogenetic precursors, including instances of imitation both of others and oneself. For Piaget, symbolic play was part of a wider, across-the-board developmental event: the emergence of the semiotic or symbolic function, which transforms the sensorimotor intelligence of infants into the symbolic, representational intelligence of children. During their second year of life, toddlers increasingly engage in actions that are only partially or schematically performed, so that they become symbols of themselves, or rather, of the complete version of the action. Piaget thought that this mechanism of progressive schematization culminated in a complete interiorization of the action and therefore of the symbol, that now becomes completely mental instead of enactive or a hybrid of exteriorized action and interiorized representation. In symbolic play children play with their ability to produce and use behaviors that work as *signifiers* of other behaviors or situations (the *signified*). In this view, imaginative play is just part of a wider function (the symbolic function) that is also manifest in behaviors such as imitation, language, drawing, and the formation of mental images.

A more recent model of symbolic play was proposed by Leslie (1987). For him the fantasy play of young children is one of the first manifestations of the ability to form metarepresentations—the same mechanism responsible for "theory of mind." Symbolic play is therefore the result of exercising a domain-specific ability to go beyond primary representations of reality (those that represent reality as it is) by creating "decoupled" representations that are kept in mind independently of reality. In this view, imaginative play emanates from the ability to understand representations as separate from reality and behavior. A child acts as if, for example, a pole is a horse because he or she can mentally pretend that the pole is a horse. In this account, the crux of symbolic play lies in the ability to understand and adopt the mental state of *pretending*: applying an inadequate representation to a situation, not by mistake but on purpose, while keeping in mind the correct representation as well.

However, it has been argued that symbolic play may occur before children are capable of fully understanding the mental state of pretending (Jarrold, Carruthers, Smith, & Boucher, 1994; Lillard, 1994; Perner, 1991; Perner, Baker, & Hutton, 1994; Smith, 2002). Ontogenetically, "acting as if" could precede the ability to understand its mental underpinnings. For example, 4-year-old children who already understand the jewel in the crown of theory of mind—false beliefs—may fail to appreciate that an alien character jumping like a rabbit cannot be pretending to be a rabbit if this character does not know what a rabbit is (Lillard, 1994). This factor may cast doubts on the attribution of symbolic and pretend intentions to younger children. In this view, the first manifestations of symbolic play are themselves the process whereby explicit symbolic representations of reality are constructed, rather than manifestations of the ability to *metarepresent* these representations.

However, it is undeniable that, independently of their cognitive underpinnings at different developmental moments, what we call symbolic play behaviors are a feature characteristic of human infants and children. From a comparative and evolutionary point of view, we can try to determine if similar behaviors appear in young monkeys and apes.

In any of the above accounts (Leslie's or Piaget's) imaginative play is an indicator of the progressive development of major cognitive processes: metarepresentations or explicit symbolic representations. This is why, apart from the theory embraced, everybody agrees that whether or not other primates show symbolic play may have crucial implications for understanding the evolution of higher cognitive processes. The diversity of views and systematizations of imaginative play makes the task of investigating its origins in nonhumans all the more difficult. Rather than using one particular set of criteria in our review of the primatological literature, we discuss those cases in which (1) authors claim to have observed imaginative play, and (2) we consider that some features of symbolic or fantasy play may be present, such as instances of incomplete actions produced (relatively) out of context, not by mistake but apparently on purpose: (a) uses of an object as if it were a different object, (b) actions that appear to take for granted an object or a situation that is not there, or (c) any other action that appears to be governed by imaginary rather than actual features of the environment.

IMAGINATIVE PLAY IN MONKEYS AND APES

Play Fighting and Chasing

One kind of play that is frequently reported in virtually all monkeys and apes is play fighting and chasing. This type of play is especially characteristic of primates, although it can also be observed in many other mammals (Fagen, 1981). Young monkeys and apes run after each other, wrestle and bite each other in a mock, playful way (see Lewis, Chapter 3, this volume and Fry, Chapter 4, this volume). We might think that this is the fantasy play of nonhuman primates—one limited to "pretending" the serious activities of fight and aggression that they may frequently observe in the adult members of their groups. Indeed, Bateson (1955) proposed that, since these playful activities of animals are accompanied by specific facial and vocal expressions (play faces and play vocalizations), they could be considered to be instances of metacommunication—a notion close to that of metarepresentation, in that both can change the literal meaning of behaviors.

There are, however, several reasons to doubt this interpretation of social play in primates. First, it emerges in young monkeys and apes that are reared in captivity without the benefit of adult models of fighting behavior. Play chasing and fighting appear to be intrinsic components of the behavioral repertoire of primates, not ontogenetically derived from their serious counterparts. Second, it would be strange that among the many adult behaviors observed and experienced by youngsters, they only playfully reproduce fight and chase patterns, ignoring mothering, feeding, sleeping, sexual intercourse, etc. Third, play chase and play fight are part of the behavioral repertoire of virtually all mammals and appear to be a phylogenetically ancient pattern of behavior that exists in its own right with its own adaptive functions (Fagen, 1981; Smith, 1982). Fourth, play chasing and fighting exist in humans alongside symbolic play, suggesting that they are distinct patterns of behavior with their own ontogenetic and phylogenetic histories.

In sum, the category of behavior we call imaginative or fantasy play appears to be distinct from so-called play fighting and chasing. The latter may have been selected phylogenetically because of their resemblance to real fighting and chasing and the practice and learning benefits this resemblance conveys, but there is no evidence that young primates are *pretending* to fight or engaging in playful imaginary chases.

Pretending as Deception in Primates

One of the terms used to refer to imaginative play is "pretending," because the psychological operation at work is supposed to consist of *pretending* that something is the case when it actually is not. In this sense, pretending is a cognitive operation that can be applied not only to imaginative play but also to deception. Play occurs when the person/animal openly signals pretense, whereas deception occurs when the person/animal tries to make others believe that the pretense is reality. Evidence of deceptive pretense could therefore be taken as an indication that primates possess the fundamental cognitive ability for imaginative play (Mitchell, 2002a).

Indeed, putative deceiving behaviors have been described in many species of primates (Byrne & Whiten, 1988; Whiten & Byrne, 1988). However, not all of them imply an element of pretense. For example, many consist of refraining from looking at a particular target or preventing others from seeing something. Suppressing cues about intentions or deeds need not involve an active attempt at creating a misrepresentation of reality. Some of the behaviors reported in primates, however, could reflect an element of pretense: for example, de Waal's (1982) observation of a chimpanzee limping only in the presence of the dominant who hurt him, but not when surrounded by others. The problem with these one-off observations is the difficulty in deciding which cognitive processes underlie the behavior. What appears to be an attempt at pretending to be injured could simply be a blind association between limping and not being threatened by the dominant acquired during the period when the subordinate was genuinely limping as a consequence of the dominant's aggression.

Whether or not that is the explanation in this specific case, the fact remains that an ability to pretend in deceptive contexts may not coincide with the ability to pretend playfully.

FANTASY PLAY IN THE WILD

The impression that imaginative play is exclusively human stems from the nearly complete absence of reports by field primatologists of instances of symbolic play. There are only a few rare reports of behaviors that bear some resemblance to imaginative play.

Goodall (1986; van Lawick-Goodall, 1968) reported one infant chimpanzee executing *in vacuo* the behavior of extending a small twig downward after seeing her mother engaging in real termite fishing from a tree. She also mentions that chimpanzee infants may engage in partially imitative nonfunctional twig manipulations while or after their mothers had engaged in termite fishing. These partial imitations may constitute a phase in the development of termite fishing skills. Something similar has been described by Matsuzawa (1999) in Taï chimpanzees that learn to crack nuts open with stones. However, neither Goodall nor Matzuzawa identify these behaviors as instances of symbolic play. Rather they present them as a stage in the acquisition of subsistence skills—one in which the chimpanzees manipulate the same objects as their elders in a serious context. If any play is involved, it is not imaginative but rather *manipulative.*

Hayaki (1985) reported his impression that a young chimpanzee was play wrestling with branches in the same way (including the emission of play pants) that chimpanzees play wrestle with conspecifics. Wrangham (1995, as cited in Parker & McKinney, 1999; Wrangham & Peterson, 1996, p. 5) observed a "lonely" 8-year-old male chimpanzee playing with a log over a 4-hour period, as follows: He carried it "on his back, on his belly, in his groin, on his shoulders," all of which are patterns of infant carrying among chimpanzee mothers. He took the log with him all the time, and when resting in his nest, "he held it above him like a mother with her baby." Moreover, at rest time he made an additional nest on which he placed the log. He repeated a similar sequence of behaviors 3 months later.

Matsuzawa (1995, as cited in Parker & McKinney, 1999, p. 132) observed a similar, though less elaborate, example. An 8-year-old female juvenile followed a mother who was carrying a sick baby. At one point, the juvenile female broke a piece of wood (50 centimeters long, 10 centimeters in diameter), put it on her shoulder, and continued to follow the mother. She then shifted the piece of wood to underneath her arm, so that she held it under her upper arm. Next, she placed the wood on a branch and "slapped it with one hand several times, as if softly slapping the back of an infant."

These observations, especially the first by Wrangham, are strongly suggestive of some primitive form of "doll play" among wild chimpanzees. The behaviors are applied to inanimate objects in ways that resemble those used by chimpanzee mothers with their offspring. As we see in the next section, the use of parenting patterns with inanimate objects is one of the candidates for symbolic play observed among captive apes.

Breuggeman (1973, p. 196) reported one instance of similar "doll play" in a 2-year-old female rhesus monkey. She moved after her mother, who was carrying a baby in a ventral position. She took a piece of coconut shell and placed it against her belly with one hand. When her mother stopped and lay on one side, resting her hand on the baby, the young female "adopted the exact posture . . . while still holding the coconut shell to her ventrum."

Zeller (2002, p. 194) reported the induction of maternal behaviors in a female Barbary macaque who was contemplating other females nursing their infants. This female, who had had no baby in the season, started to hold and carry a 2-year-old younger sister "as if she were an infant[; she] cuddled her, groomed her, held her to the ventrum and lip-smacked to her in the pattern of the other young females with their infants." She acted in this way "for a number of weeks." Zeller suggests that this female might have been pretending that she had an infant like the others. Practically all the (rare) instances of potential pretend play in wild primates appear to involve the application of maternal behaviors to inanimate objects or inappropriate animals. Parker and McKinney (1999) state that these observations suggest that symbolic play is not exclusive to captive, "enculturated" apes, but a natural chimpanzee behavior. However, the problem remains of why it is such an infrequent behavior that has only been observed on a few occasions and does not appear to be part of the usual repertoire of any particular chimpanzee community, not even of any individual chimpanzee? At the very least, we should conclude that symbolic play is not a characteristic pattern of wild chimpanzees, but a marginal, exceptional occurrence, in contrast to human beings in whom symbolic play appears to be a universal feature across cultures. (In some human communities symbolic play may be relatively infrequent, but some imaginative play patterns are discernible; see Smith, Chapter 8, this volume; e.g., the Pasiegos, a rural community of Northern Spain whose economy is almost exclusively dependent on milk cows, have a traditional toy, the "cow," made with a V-shaped branch, and their children are capable of engaging in imaginative play activities when induced to do so: Martín-Andrade & Linaza, unpublished observations).

FANTASY PLAY IN COLONIES OF CAPTIVE PRIMATES

De Waal (1982) reported two instances of young chimpanzees copying the funny way of walking by adult chimpanzees affected by a permanent deformity or a transitory injury. In the first case, the young chimpanzees imitated the "hunched-up way of walking" of a crippled female during several days: "they would walk behind her on single file all with the same pathetic carriage" as she demonstrated (p. 80). In the second case, after severe fighting among adult males, one of them had to walk on his wrist instead of his hand, which provoked a bizarre asymmetrical walking. All the young chimpanzees would "imitate him and suddenly begin stumbling around on their wrists" (p. 135).

What is remarkable about these observations is that the young chimpanzees imitate unusual patterns of walking, which the adult models cannot prevent themselves from performing, but that the youngsters must be performing on purpose. Since de Waal is not very explicit about the accompanying behaviors (e.g., we do not know if the young chimps were showing play faces), it is difficult to decide if these imitations are simple instances of symbolic play.

De Waal (1986) has also reported several instances of possible pretense involving the simulative reproduction of behaviors in deceptive contexts; for example, the case already mentioned, of the adult male who feigned a limping gait when in view of the dominant.

FANTASY PLAY IN HAND-REARED APES

Simply rearing an ape in a human household environment does not appear to promote more instances of symbolic play. Kellogg and Kellogg (1933) raised their son with a chimpanzee "sibling" during his first year and a half of life, in order to conduct a strict comparative study of development; they comment that, in contrast to their son, the chimpanzee infant engaged in very little "imitative play." A similar conclusion was reached by Ladygina-Kohts (1935/2002) when comparing her son with a chimpanzee baby she had reared several years earlier. The chimpanzee's play lacked imagination and pretense. The only instances of possible pretending occurred in deceptive contexts, although it was reported that instances of imitation in the handling of objects that did not appear to pursue a goal (at least, not successfully; e.g., clumsily handling cleaning implements), and a few instances of rough-and-tumble play in which objects were included in such a way that the objects appeared to take the place of a missing partner. Nevertheless, her overall conclusion was that, in comparison with a human child, imagination and fantasy were lacking in the chimpanzee.

Mignault (1983, 1985) found very little in the way of imaginative or symbolic play in four nursery-reared chimpanzees who were intended to be reared "in an environment similar to that of human children," in that the subjects had access to humans every day and to "many objects and toys" (1985, p. 751). The chimpanzees could also observe some social routines performed by humans, "for example, cooking" (1985, p. 751). When tested with a set of objects that was previously used in studies with human children of similar ages (Inhelder, Lézine, Sinclair, & Stambak, 1972) and was designed to capture transitional behaviors between Piagetian sensorimotor intelligence and symbolization, the chimpanzees showed very few of such behaviors. Those reported included one chimpanzee "scribbling" on a piece of paper with a spoon, on one occasion, with a comb, on another. Mignault (1983) points out that it looked more as if she were checking to see if she could produce marks with these objects, because she immediately abandoned her efforts.

One of the sharper differences between chimpanzees and humans occurred in their treatment of dolls. Children used dolls as recipients of actions such as cleaning and feeding, whereas the chimpanzees tended to treat dolls as any other object, frequently breaking them (Mignault, 1983). To encourage the chimpanzees, they were presented with models of pretend behaviors, such as a cleaning sequence or a fictitious child care sequence acted out with the dolls. Mignault (1983) found some deferred imitation (imitating some conventional use of the objects demonstrated; e.g., one instance of applying a brush on the doll, but not on the hair, as demonstrated), but in comparison with children the chimpanzees did not show any major trend toward adopting a conventional use of objects—one of the first manifestations of imaginative play, according to Inhelder and colleagues' (1972) analysis—let alone toward producing instances of fully symbolic behavior.

PRECURSORS OF SYMBOLIC PLAY IN HAND-REARED GORILLAS

In a longitudinal study of hand-reared gorillas (Gómez & Martín-Andrade, 2002) we identified several instances of behaviors that resemble simple cases of symbolic play in humans. We suggested that these are "precursors" rather than full-fledged instances of symbolic play. We classified these precursors into categories corresponding to types of symbolic play in children: *object substitution* (such as transporting and handling inanimate objects [e.g., stones, shoes, balls], using patterns reminiscent of gorilla maternal behavior); *context substitution* (building incomplete nests similar to those used for rest during episodes of mutual vigilance in social play); *imaginary properties* (imitating cleaning with a cloth when no dirt remains to be cleaned, or drinking from an empty cup); *role enactment* (offering reinforcements to the trainer during learning experiments). We argued that all of these instances could be explained without the attribution of a pretend intent to the gorillas (see discussion for details of our arguments).

FANTASY PLAY IN "LINGUISTIC" AND "ENCULTURATED" APES

If observations of potential cases of imaginative play are scarce in the wild or in captive animals housed in traditional zoo conditions or hand-reared by humans, the situation changes, to some extent, among apes that have been subject to special symbolic training in rich environmental conditions.

"Linguistic" apes are those that have been subjected to a systematic training regime to try to teach them the use of artificial signs or symbols like those of human language (Gardner & Gardner, 1969; Premack, 1976; Rumbaugh, 1977). Premack (1983) argued that this linguistic training regime may produce changes in the cognitive processes of apes. Specifically, he believed that the symbolic codes learned allowed them to successfully solve abstract tasks, such as those involving comparing relations between relations (e.g., understanding that the relation between two identical objects [e.g., two cups] is the same as that between a different pair of identical objects [e.g., two spoons]).

"Enculturated" apes are those that have been reared by humans in a largely human environment, both physically and socially. This label was created by Tomasello, Kruger, and Ratner (1993) and reflects the idea that in such rearing conditions, the cognitive processes of apes may be modified by the social shaping that humans may exert upon their patterns of attention and object manipulation. All linguistically trained apes have been enculturated, but not all enculturated apes have been linguistically trained. For example, some of the hand-reared apes we discussed in the previous sections might be considered to be enculturated, to some extent. However, it is difficult to establish objective criteria to decide what qualifies as enculturation. Many times detailed accounts of the patterns of interaction between apes and humans during the hand-rearing period simply are not available.

Some authors reject the notion of enculturation as a process that can modify apes' cognitive processes. They believe that enculturated apes are simply given the opportunity to develop in a rich and stimulating environment and therefore have been prevented from following an abnormally understimulated pattern of development, as would happen to most captive apes confined in cages. The minds of enculturated apes would then be just normal ape minds, comparable to those of wild animals (Mitchell, 2002a; Parker & McKinney, 1999).

Linguistic and enculturated apes are the source of the most systematic and rich descriptions of instances of fantasy play.

Viki's Imaginary Pull-Toy

This account of Viki is the most famous single observation of an episode of fantasy play in a nonhuman primate. Viki was a chimpanzee hand-reared by C. and K. Hayes, with the intention of conducting a study similar to that of the Kelloggs's: to expose her to the same kinds of experiences as human children. To their domestic rearing the Hayeses added systematic attempts at teaching spoken English to Viki. These were mostly unsuccessful, although they report that Viki did manage to produce a few spoken words with some degree of success (Hayes, 1951).

When Viki was 16 months old, during 3 consecutive days, Cathy Hayes (1951, pp. 80–84) observed her "slowly and deliberately . . . marching around the toilet trailing the fingertips of one hand on the floor"; she would pause now and then, glancing back at the hand, and then resume her walk. Hayes reports that her first impression was that Viki was enjoying the vibration of her fingertips on the linoleum, but she suddenly realized that the posture used by Viki was the same she used to drag pull-toys behind her. She clarifies that Viki was at the "pull-toy stage"—she enjoyed "dragging wagons, shoes, dolls, or purses," and she did so by adopting exactly the same angle and posture: Walking with the legs and one hand, she used the other hand to pull the string of the toy. Hayes thought that Viki had "an imaginary pull-toy." After this insight, Hayes observed the following variations in successive days.

One day Viki interrupted her action, turned and made "a series of tugging motions," then moved her hands over and around a plumbing knob, placed her fists "one above the other in line with the knob and strained backwards as if tugging, until after a little jerk she resumed her trailing motion" (Hayes does not clarify if Viki had experienced this sort of accident in her real manipulations of pull-toys). Another day, standing on her potty (for some reason this game always took place in the bathroom) Viki "began to raise the 'pull-toy' hand over hand by its invisible rope." This was again reminiscent of one of her favorite games with real pull-toys: raising them from the floor while standing on a piece of furniture. Moreover, after that, Viki "lowered it gently and 'fished' it up again."

The trailing game became more and more frequent, although always confined to the bathroom area. Another day, after the game had been going on for about 1 month (Hayes only informs us that the game started in January and that this incident happened early in February), Viki repeated the plumbing knob routine (what Hayes describes as pretending that the toy was stuck), but this time she stopped her pulling, turned to Cathy Hayes and called her "Mama, Mama," using one of the few recognizable words of her vocabulary. She did this when she needed help. Hayes believed she was asking her to join in her pretense, and so she did: She pretended to untie the tangled string, and she handed it back to Viki, who then resumed her trailing at an unusually fast pace.

Hayes then decided to produce her own imaginary pull-toy performances in front of Viki, with the added improvement of sound effects accompanying her trailing movements. She reports that, watching her pretense, Viki approached the area where the imaginary pull-toy should be, "stared transfixed [it is not clear if she stared in the direction of the imaginary pull-toy] and then uttered her awstruck 'boo' " (p. 83). Unfortunately, the next time Cathy Hayes tried her own version of the imaginary pull-toy, Viki reacted with distress and jumped into her arms, and from that day Viki never again played the game.

This was the only case of pretend play reported about Viki during the entire study, which lasted for almost 7 years. Some photographs in Hayes's book, though, depict Viki handling a mop and doing "sewing," activities that presumably she was not carrying out seriously. Hayes and Hayes (1952, p. 451) indeed report a spontaneous tendency in Viki to engage in "playful imitation" of activities, such as "dusting furniture and washing clothes and dishes." It is unclear if Viki's actions would qualify as symbolic play, because she tended to use the real objects for her imitations.

Adopting the skeptical stance, we could point out that the only report of this pretend incident has been published in a popular book with an informal description. Moreover, the initial report by Hayes describes the chimpanzee trailing her own finger along the linoleum. Only after the idea suddenly occurred to Hayes that Viki was playing with an "imaginary pull-toy" did the description turn into an activity of pulling rather than dragging. However, the subsequent elaborations described by Hayes appear to go well beyond the pattern of dragging fingers. Furthermore, the fact that this series of episodes occurred during the period when Viki was actively dragging real pull-toys and objects suggests that she might have been genuinely reproducing the dragging pattern *in vacuo.* If the elaborations described by Hayes are an accurate description of what Viki did and not a licence of the popular style of the book, they appear to involve actions with objects that are not really present and therefore could be best described as "imagined."

Imaginative Play in Signing Chimpanzees

Gardner and Gardner (1969) report instances of routines such as bathing performed on dolls by Washoe, the sign-language-trained chimpanzee, when her estimated age was 18–24 months. The Gardners had bathed her regularly, following a well-established routine from the beginning of the project. They also provided her with dolls. In the 10th month of the project, "[Washoe] bathed one of her dolls in the same way we usually bathed her. She filled her little bathtub with water, dunked the doll in the tub, then took it out and dried it with a towel" (p. 666). They report that this routine, which sometimes included soaping the doll and, other times, more fragmented actions, was repeated "many times" (p. 666). In other reports, they comment that Washoe kissed and fed dolls and liked "to wash the dishes, in a childish way" (Gardner & Gardner, 1985, p. 159) and "to set and clear the table . . . in a childish way" (Gardner, Gardner, & Van Cantfort, 1989, p. 1). Unfortunately, no details are provided of these activities or the contexts in which they first appeared and were maintained.

More detailed descriptions are given by Jensvold and Fouts (1993) in a study specifically intended to address the issue of symbolic play in Washoe and other members of a colony of signing chimpanzees. The researchers videotaped 15 hours of behavior of these linguistically trained chimpanzees after providing them with a number of objects and leaving them on their own. They observed six potential instances of imaginary play. In one, while manipulating a stuffed bear, the chimpanzee Dar signed "tickle, tickle" on the bear (by which the authors presumably mean that instead of touching the back of her own hand with the index finger—the typical shape of this sign—she might have touched the bear; this is unclear in their description), and then pushed the bear between her shoulder and the fence. According to the authors, Dar usually pressed herself against the fence when tickled by a human. Their interpretation is that Dar was inviting the bear to tickle her—which, if true, would be an instance of animation attribution to the bear.

Dar produced a second example of animation, presumably with the same bear, 6 minutes later. After "pocketing" it in her pelvis she looked away and signed "peekaboo." The authors suggest that she was pretending to play peekaboo with the bear.

A few days later, Dar signed "tickle" twice on a puppet, and then she looked away. The authors consider this example as one more instance of symbolic play with animation attribution. They also classify as animation pretense Washoe's behavior of "hitting away," with a play face, at a stuffed bear.

The authors also report two possible instances of object substitution. In one, Washoe pressed a brush with her arm against her body and signed "book in." The authors clarify that Washoe often placed books in that way, between arm and body. In the second instance, Moja placed a shoe on one of her feet and a purse on the other, and later replaced the shoe with the purse. She signed "shoe" while looking at the shoe. The authors suggest that she was pretending the purse was a shoe.

Finally, Jensvold and Fouts (1993) report a further observation with Washoe, in which she picked up a toothbrush, signed "hairbrush," and then used the toothbrush to brush her hair. As to the paucity of examples recorded during these 15 hours of videotape, the authors point out that in some studies with children, similarly low frequencies have been reported.

Lexigram-Using Chimpanzees

Austin, a chimpanzee trained to use lexigrams (an artificial, computerized language based on abstract geometric designs: Rumbaugh, 1977), was reported to pretend "to eat imaginary food, sometimes using an imaginary plate and an imaginary spoon" (Savage-Rumbaugh & Lewin, 1994, p. 277). No description is given of the precise actions performed by Austin with the "spoon" and the "plate," but he is described as carefully placing "the nonexistent food in his mouth and to roll it around in his lips watching it just as though it were real food." Sherman, another chimpanzee reared and trained with Austin, used dolls, "particularly King Kong dolls," to pretend that they "were biting his fingers and toys as well as having fights with each other" (Savage-Rumbaugh & Lewin, 1994, p. 277).

Lastly, these authors report what they believe was an episode of joint pretense involving Austin and Sherman as they were watching a King Kong movie on TV. When Kong was placed in a cage similar to one they had in their room, they stopped watching the movie, turned to the real cage and started to "make threat barks at the empty cage and to throw things at it" (p. 277). One of them, Sherman, "got out the hose and began to spray the cage." The authors suggest that they were pretending that King Kong was in the cage.

Sarah: Chip Language

Premack and Premack (1983) briefly report a possible episode of symbolic play in Sarah, a linguistically trained chimpanzee (with plastic chips instead of signs). She had been repeatedly given a "face puzzle" (a photo of a chimpanzee face, cut into pieces) as part of a test of her ability to reconstruct a whole from its parts. One day, after watching herself in a mirror trying on real hats, she started to arrange the elements of the face puzzle in a new way. She turned the nose and mouth pieces over, placing them on the head of the photo, over the eyes, like a hat; in another occasion she placed the peel of a banana in a similar position on the photo. She had never done this before her hat-wearing experience with the mirror, and she ceased to put "hats" on the photos when she no longer was given real hats with which to play.

PRETEND PLAY IN KOKO, THE LANGUAGE-TRAINED GORILLA

Patterson and Linden (1981) and Patterson (1978, 1980) informally reported some instances of possible pretend play in the sign-trained gorilla Koko. Matevia, Patterson, and Hillix (2002) review the evidence for pretend play in this gorilla in a more systematic way. One problem with examples of pretense in nonlinguistic apes is the need to infer the pretend intent from actions alone. In children there are also many behaviors that are difficult to classify as involving pretense. This ambiguity disappears or is reduced when they use language to identify the action they are doing as "drinking," "sleeping," or "cleaning," or even more decisively, when at an older age they proclaim they are "just pretending."

One advantage of linguistic apes is that they might be able to clarify what they are doing with their available symbols. Indeed, this is the clarity that Matevia, Patterson, and Hillix (2002) claim to have found with Koko.

Koko's Doll Play

One of the earliest and more characteristic manifestations of pretense in children is when they use dolls as actors or recipients of some of the actions of their everyday lives. Koko liked dolls and engaged in a variety of doll manipulations.

At 5 years, Koko was observed placing two toy gorillas before her and signing "bad, bad" while looking at one of them (a sign her caretakers frequently used to reprimand her) and then "kiss" while looking at the other. She then signed "chase tickle"—the signs she uses to request some of her favorite play interactions—and hit the two toy gorillas together, in what could be a pretense of that kind of play. Immediately, she started to "tussle" herself with the two dolls (Matevia et al., 2002, p. 288), or, as rendered in what appears to be an earlier description of the same observation, Koko "joined in and wrestled them both at once" (Patterson, 1980, p. 541). When she ceased to play, Koko signed "good gorilla good good." Patterson (1980) comments that she observed similar behaviors at that time, but that this doll play tended to occur when Koko was alone, and she would interrupt her playful activity if she discovered that humans were observing her.

Furthermore, Matevia and colleagues (2002) report instances of Koko (at 12 years of age) placing a small doll in the mouth of another doll, which they believe may imply some simulation of biting, given the contextual comments by Koko, such as "that trouble there," pointing to a smurf in the mouth of a dinosaur (although, in this particular example, the pretend scene had been suggested through comments by one of the human experimenters).

Koko also kisses dolls while making the sign "good" or places them one against another in such a way that the human experimenters report that she is making them kiss each other. In another observation Koko placed a doll against her nipple with one hand while signing "drink" with the other. She also is described as cradling dolls against her chest.

A more complex example, recorded at 16 years of age, involved Koko placing her finger into a toy lizard's mouth and then withdrawing it and shaking her hand. When asked with signs if the alligator had bitten her, she repeated the sign "bite."

At 19 years of age, Koko, after receiving an orangutan doll as a present, and after signing "drink," was asked by Penny Patterson "Where does the baby drink?" using both speech and signs. Koko responded by taking the hands of the toy orangutan and placing the latter's thumb into the doll's mouth (similar to the sign "drink") and immediately placing the doll's index finger on the mouth, forming (according to the observers' interpretation) the sign for mouth.

Another elaborate example reported by Patterson (1978) occurred when Koko received a set of toy teacups. First she is reported to pretend to drink from the empty cups and to encourage Patterson to do so by signing "sip" (a sign Koko invented by blending together "drink"and "eat," and which, according to Patterson, Koko uses specially in pretend situations), and later by gently nudging the hand with which Patterson was holding the cup to her lips. Later Koko was asked by Penny to offer drink to a chimpanzee doll, and the gorilla did so at the same time that she signed "drink" while looking at the doll.

Koko's Object Substitution

On one occasion, Koko removed the label off a bottle of water, placed it on her head, and signed "hat" (a sign that incidentally consists of tapping one's own head). On another occasion, she signed "hat" when a human experimenter placed a strawberry on a toy alligator's head.

Other reported instances of object substitution are more controversial. For example, Matevia and colleagues (2002, pp. 195–196) describe how Koko uses a piece of chewing gum in a variety of ways: chewing, poking it against a molar, placing it in a toy's mouth, or placing it up her nose and blowing it out. When Koko placed the gum in the toy's mouth, she signed "lip"; when she returned it to her own mouth, she signed "clay"; after rubbing it back and forth on her mouth, she signed "lipstick"; and finally, when blowing it out of her nose, she made no sign. Based on these signs and the context, the authors suggest that Koko uses the piece of gum as a substitute for a variety of objects, namely, lipstick, a tooth filling, food for the toy, and nasal mucus. It is, however, not always clear why the authors suggest these interpretations. Maybe Koko was simply exploring the possible manipulations of an object in relation to herself and other objects.

Koko's Role Playing?

Matevia and colleagues (2002) suggest that Koko might be able to pretend to be an elephant. One day, in response to being presented with a long rubber tube, Koko signed "sad elephant." Seeing the rubber tube may have triggered in her the association with the idea of an elephant (why she used "sad" is less clear, although she might have felt frustrated because her caretaker had been refusing to give her fruit juice). The caretaker asked her, with signs, if she was a sad elephant (it is unexplained why the caretaker suggested this interpretation), to which Koko replied, using exactly the same signs: "sad elephant me." The caretaker then added, with signs: "I thought you were a gorilla," to which Koko replied, "elephant gorilla thirsty." It is very difficult to evaluate what the gorilla is trying to say (at least, for the naive reader), but it seems that the pretense element was introduced and maintained by the caretaker.

Something similar occurs in another exchange with the same caretaker. She showed Koko a picture of a bird in a magazine. Koko made the signs "that me" (which essentially consist of pointing to the magazine and then pointing to herself), which the caretaker immediately interpreted as meaning that Koko was pretending to be a bird. This gave rise to a dialogue, in which Koko used the signs "Koko" and "bird" together a couple of times. The authors clarify that, in contrast to what occurs when children pretend to be an animal, Koko never accompanied her supposed pretend claims with any action depicting a bird (not even when she was invited to fly, to which she responded with the signs "fake bird clown").

This example illustrates the difficulty of using Koko's signing as a source of clarification for her possible imaginative play. In the absence of a systematic characterization of the way in which she uses her signs, it is difficult to understand why "that me" should mean "I am that" instead of "Give that to me" or "I like that," "Is that for me?", or "What do you expect me to do with that?" Similarly, in the absence of a description of how the word "fake" was taught to Koko, it is difficult to assess the significance of her response to the invitation to fly.

Indeed, Koko is reported to have used the term "fake" or "faketooth" on other occasions (apparently she is obsessed with the dental features of visitors, and she uses that sign when she discovers dental prostheses in her visitors' mouths). For example, in one occasion she reacted with fear when a caretaker produced a new toy dinosaur with a gaping mouth (presumably, covered in teeth). When the caretaker asked her what it was, she replied "faketooth fake." Koko uses "toothfake" to refer to new visitors who have false teeth; perhaps she was identifying this new toy, with its presumably conspicuous false teeth, as a "faketooth."

Another interesting aspect of Koko's behavior is her use of gorilla laughter and smiles during potential episodes of pretense (Patterson, 1978), although no systematic report exists on this area.

Koko's Attribution of Objects and Properties

Koko is described as engaging in actions such as the following: lifting an empty toy cup to her lips and making "loud slurping sounds," as if she were drinking, or lifting an empty bowl and pretending that she is eating. She also places an empty bowl on a toy cat and takes the cup to her own lips and makes slurping noises. In this same example, she then put a sock on the toy's head and signed "hat," wrapped a cloth around the cat, and finally kissed it.

Finally, Matevia and colleagues (2002, pp. 300–301) report one possible instance of Koko's engagement in cooperative storytelling, when one of her caretakers improvised a signed tale and prompted Koko to provide input from time to time. For example, she asked Koko what she wanted the story to be about, and Koko made the sign for "alligator." The caretaker started a story about an alligator who went in search of ice cream and on his way found . . . and here she again asked Koko to contribute by asking her "what?." Koko responded with the sign for "bird." The caretaker elaborated the story, explaining that the bird rode on the alligator; "where on?" she asked Koko; "nose" was Koko's response; and so on. In the absence of further information, it is again difficult to assess the significance of this interesting interaction with signs. Did Koko really understand the story line and produce her responsive signs with a sense of the imaginary context (apparently no props were used to support the signed narrative), or were her responses specific to each question and put together only through her caretakers' skill?

Koko might be the "linguistic" ape that produced the most apparent examples of imaginative play. We have, however, no direct estimate of the frequency of these pretend examples. As with other "linguistic" apes, many of these tantalizing descriptions are difficult to evaluate without more details about their context and history.

FANTASY PLAY IN BONOBOS

Kanzi is a bonobo that has been taught an artificial symbolic language made of external geometric designs (lexigrams) that he can either point to or press on a computer (Savage-Rumbaugh & Lewin, 1994). His teaching experience consisted of informal but intensive interactions with people who demonstrated the use of the artificial language in many different routines designed to reproduce situations in which human children learn their linguistic skills. Kanzi has reportedly learned to understand spoken sentences in English (see Savage-Rumbaugh, Shanker, & Taylor, 1998).

Kanzi has also been reported to engage in pretend play activities. His favorite is the handling of entirely imaginary food: "He pretends to eat food that is not really there, to feed others imaginary food, to hide such food, to find it, to take it from other individuals, to give it back to them, and to play chase and keep away with an imaginary morsel" (Savage-Rumbaugh et al., 1998, p. 59). Perhaps the peak of his imaginary actions occurs when he puts the imaginary food on the floor and then acts "as if he doesn't notice it" until one of the persons pretends that she is reaching for the "food," at which point Kanzi "grabs it before they can get it" (p. 59).

Interestingly, Savage-Rumbaugh and colleagues (1998) report that such games start when Kanzi pretends to take and eat (or to take and give to someone) a piece of food depicted on a picture or a TV screen. (The same imaginary snatching of food from pictures is reported for Panbanisha, Kanzi's mother, in Savage-Rumbaugh & Lewin, 1994). Kanzi is also reported to feed imaginary food to toy dogs. He "engages in imaginary games with toy dogs and chimpanzees, as children play with dolls" (Savage-Rumbaugh et al., 1998, p. 60). These games appear to consist of the following: pretending that "a toy dog or toy gorilla is biting him or is chasing and biting a person" (p. 60). He might pretend to be scared when chased by

Savage-Rumbaugh, wearing a monster mask. If the latter stops the chase, he doesn't give evidence of fear but "seems puzzled," according to her.

His younger sister, Tamuli, also plays with chimpanzee dolls or substitutes, such as a dead squirrel. Savage-Rumbaugh describes the dead squirrel episode in some detail: Tamuli carried the dead squirrel "as if it were a baby . . . [pulling] its little feet around her waist as though it were clinging to her. She groomed it . . . and pretended to nurse it" (p. 61). Tamuli had just had a younger sister and therefore had been exposed to the nursing and mothering behaviors of their mother. According to Savage-Rumbaugh and colleagues, Tamuli took the dead squirrel with her, following her mother, "doing everything to the squirrel that she saw her mother do" with her sister (1998, p. 61). When the squirrel was taken away from her, she was sad and depressed the rest of the day.

Note that this reaction of sadness and, indeed, the whole episode of "pretense" is ambiguous in interpretation: On the one hand, it could be due to the disappearance of a favorite toy especially suited for pretend; on the other, it could reflect a real application of mothering behaviors to the squirrel (as wild and captive apes have been frequently observed to do with a variety of objects, including dead infants). After all, if pretense were the essence of the game, any other substitute, for example, one of their ape dolls, would have sufficed.

FANTASY PLAY IN ORANGUTANS

Little information is available about imaginative and pretending behaviors in orangutans. Miles (1991) reports that Chantek, an orangutan trained in sign language, "signed to his toys and offered them food and drink. . . . He also appeared to express sympathy for his toys by rescuing them from caregivers' apparent threats" (p. 16). The action of feeding a toy animal was first seen when Chantek was 20 months old (Miles, Mitchell, & Harper, 1996). He is also reported to have acted "afraid of a nonexistent cat" at 25 months (p. 284) and to imitate caregivers' behaviors, such as cooking in a nonfunctional way (e.g., putting cereal in a pan, and the pan on a switched-off stove); in fact, one such imitation involved an unfortunate episode of object substitution: Instead of cereal, Chantek placed a video camera in a pan with water.

Russon (2002) discusses four cases of possible pretense with a deceptive intent, such as acting uninterested in rehabilitant orangutans. In one of them, one adult female (apparently non-enculturated) focused her attention on a human using a Swiss army knife to cut and strip bark off a branch.

However (as revealed by a detailed analysis of videotape footage), her interest was eventually caught by the unguarded backpack of this person. Russon describes how the orangutan resumed her focus on the cutting demonstrations of the human but slowly changed her posture until she was in a better position to make an attempt at catching the backpack. At that moment, she abruptly abandoned her interest in the cutting and dashed for the backpack. Russon discusses to what extent this behavior involved an ability to feign interest in a false goal and therefore some ability to pretend what is not the case. However, we suggest that more relevant to the issue of symbolic play might be the reaction of this female orangutan when she first saw the cutting and stripping of the branch: "she reproduced the sawing and scraping using one stick against another" (p. 231). This imitative reproduction of the action, which is very similar to some of the earliest manifestations of pretending in humans, involves the use of a substitute object. It is unclear what its function is, apart from making explicit (to the orangutan) what she found of interest in the event.

EYE-COVERING GAMES IN PRIMATES

A popular type of play among many monkeys and apes consists of walking and running around with their eyes closed or otherwise covered (for a short critical review, see Russon, Vasey, & Gauthier, 2002). For example, to quote our own personal observations, young hand-reared gorillas run around with their eyes tightly closed or covered with a piece of cloth, stumbling against objects and structures in the environment and exhibiting exaggerated movements typical of motor play and signals such as play faces and vocalizations.

It has been suggested that eye-covering play may involve pretense; primates might be pretending that they cannot see. More conservative interpretations suggest that it is a form of self-stimulation or action play (Gautier-Hion, 1971, as cited in Russon et al., 2002). On the basis of a systematic study with orangutans and Japanese macaques, Russon and colleagues (2002) propose that although it does not appear to involve pretense, eye-covering play may involve engaging in a "cognitive game similar to pretending—playing with the discrepancies among multiple mental images of a situation" (p. 254). They observed that captive Japanese macaques and, more frequently, rehabilitant orangutans engage in "quiet" varieties of eye-covering games. For example, the primates would walk slowly along a route with their eyes closed or covered and without producing exuberant, playful movements or play signals. The orangutans would fixate a distant goal and then slowly approach it with their eyes covered. At difficult points involving obstacles, they would peek briefly and then resume their blind walk, or grope with their hands as if in search of an impending obstacle.

Russon and colleagues (2002) argue that these instances of eye covering cannot be explained as self-stimulating or sensorimotor play. They propose that eye covering involves *imagination* (in the sense that the behavior of the primates is guided by stored visual representations rather than online visual perceptions), and that this would involve some form of *secondary representation* (because the subjects seem to be contrasting potentially discrepant visual and tactual representations). They speculate that, in the case of the orangutans, some form of *metarepresentation* might be involved (because they are able to plan in advance and execute complex routes). This ability to play with the contrast between the stored visual representation and the incoming tactual information might involve some precursor to the ability to simultaneously hold two contrasting representations of a situation that is involved in pretense. However, since one representation is not purposefully distorted to make it differ from reality, they conclude that no pretending is involved.

Indeed, the visual representation and the incoming tactual–kinesthetic information are not different but complementary representations of the same situation, and the aim of the primates appears to be to make them match (e.g., by peering or groping when they detect or anticipate a problem), rather than to purposefully maintain a representational discrepancy. In this sense, eye-covering games appear to be very different from imaginative or fantasy play.

Russon and colleagues' (2002) observations, however, suggest that macaques and orangutans are indeed using representations of the environment to guide their behavior (suggested by their anticipation of obstacles they cannot see), and that these are flexible enough to be separated into components, which might reflect an ability to handle those representations explicitly in their minds—a skill that might contribute to symbolic play. Indeed the orangutans demonstrated an ability to "imagine" the scenes they purposefully chose not to see. But it is unclear that this is an instance of imaginative play, as the imaginative component is not used to go beyond reality but rather to replace the sensory information that they choose not to receive online.

Finally, an unexplored angle is the question of the extent to which this type of play derives from what might be a common experience of primates: being in the dark during obscure nights. Maybe eye-covering primates are reenacting, out of context, patterns of blind locomotion that are well known to them. In fact, this would make eye-covering games more similar to early symbolic play that is based on the reproduction of action schemas out of context. A second source of common experience might be the fact that during very intense play bouts (including social play such as wrestling) primates may close their eyes in their ludic frenzy. This explanation would apply to sensorimotor and self-stimulating instances of eye covering, but not to the quiet versions described by Russon and colleagues (2002).

In sum, Russon and colleagues (2002) might be right that eye-covering games reflect an ability for imagination in the sense of generating and keeping in mind representations without the direct support of sensory information (although the tactual–kinesthetic information may act as an indirect support of the "imagined" visual representation). More importantly, eye-covering games may reflect an ability to keep perceptual–motor representations in mind in an explicit way, and this may bear some resemblance to some of the functions subserved by symbolic play—although, in themselves, these games are not instances of imaginative play.

DISCUSSION

It is difficult to evaluate the evidence about fantasy play in apes, because the available descriptions frequently omit crucial information about the historical background, context, and details of the behaviors described and interpreted as imaginative. This omission adds to the intrinsic difficulty of identifying symbolic play unambiguously. Many descriptions of symbolic play in nonhuman primates are therefore vulnerable to the adoption of a skeptical stance. To be fair, a similar vulnerability could affect many descriptions of imaginative play in human infants and children, if they were considered on their own merits.

All in all, however, it is possible to discern a pattern in the kind of fantasy candidates independently reported by different observers of nonhuman primates. First, instances of possible imaginary play are *rare* among nonhuman primates. They are not a characteristic behavior pattern of monkeys and apes, in contrast to human children. Even in the case of symbolically trained apes, imaginative play does not appear to be the pervasive, all-encompassing phenomenon that dominates human infancy.

Second, among the rare instances of possible fantasy play, a few patterns appear to be dominant. A subset of maternal behaviors—mostly those corresponding to holding and transporting infants—can be applied to inanimate objects such as logs in the wild or bundles of straw, shoes, rubber balls, stones, dolls, dead animals, and so on, in captivity. This behavior could constitute an instance of imaginary play with substitute objects. In our analysis of some instances of this behavior among hand-reared gorillas (Gómez & Martín-Andrade, 2002), we argued that a more plausible explanation would be that they emerge out of basic "instinctive" behavior patterns designed to catalyze maternal behavior that are accidentally applied to inanimate objects. These behavior patterns are normally activated by real babies, whose reactions (suckling, whining) contribute to the differentiation of these patterns into elaborate maternal behaviors. Perhaps our captive gorillas started to apply these basic patterns to inanimate objects in the absence of any direct experience with real babies or any vicarious experience of real gorilla mothers nursing and caring for their babies. Their only source of information on maternal behavior would have been their own experience as babies during the first few months of their lives. Indeed, our subjects were wild born, captured by furtive hunters. Their exact age at the time of their capture is unknown, but could be around 4–8 months. This time span could have been sufficient to accumulate experience about maternal behaviors. But the enigma remains of why their mother-playing games only appeared at around 2.5 years of age, 2 years after they had experienced the models.

The best explanation of our observations of gorilla behaviors appears to be that they are accidental applications to inanimate objects of patterns that were evolutionarily designed to catalyze gorilla maternal behavior, without any symbolic or imaginative component—there is no evidence that the gorillas were representing the objects as babies. We suggested that "a more convincing case for symbolic play would be made if this sort of proto-maternal behavior were observed in individuals shortly after they had witnessed actual maternal behaviors" (Gómez & Martín-Andrade, 2002, p. 260). Zeller (2002) and Matsuzawa (1999) report precisely this case. However, Zeller's macaque—a young female who failed to have her own baby while her "friends" had theirs—applied the maternal patterns to a younger sister rather than an inaminate object. No pretend or imagination appears to be necessary here. Matsuzawa's observation of a young chimpanzee is more convincing, although the pretend element—that the chimpanzee is representing the log as a baby—is, as usual, difficult to prove.

Finally, Wrangham's observation of a similar behavior in one chimpanzee offers the novelty that the player is a young male, which might make our explanation of accidental use of instinctive patterns more unlikely. However, in the absence of information about the history of this individual (for example, to what extent he had played with young chimpanzees, holding and transporting them in similar ways), it is difficult to draw firm conclusions about this example. Perhaps the chimpanzee was activating well-known behavior patterns already in his repertoire but this time applied to an unusual object. This kind of object substitution might be better described as "using instead" rather than "using as if": That is, the chimpanzee need not evoke the representation of a baby while performing his actions with the log (Gómez & Martín-Andrade, 2002).

We would like to emphasize that these criticisms, raised from a "skeptical stance," are also applicable to most instances of early symbolic play in human infants. It is unclear to what extent human infants are pretending when they first "feed" or cuddle a doll. Certainly there is currently robust evidence that they are not "pretending" in the strong, metarepresentational sense of the word. It is even unclear to what extent they understand the symbolic nature of their actions (e.g., that their feeding of a doll with a toy spoon is a model of their own feeding by their parents with real food). The challenge would remain of explaining how and why evolution has rendered human infants and children so prolific at producing these proto-pretending behaviors, whereas they remain so rare among our closest relatives.

ENHANCED IMAGINATION IN SYMBOLIC APES?

Reports of possible symbolic and imaginative play are more frequent and complex in apes that have been linguistically trained in a humanized environment. We do not know exactly how frequently or routinely these behaviors occur, and their level of complexity is frequently open to question due to a lack of background information about the individual history of the behaviors. The convergence of independent descriptions suggests, nonetheless, that would-be imaginative behaviors are more frequent in these than in wild or untrained captive apes, and a few patterns stand out as typical: namely, the use of dolls to perform some well-known actions with them, such as play wrestling or feeding, and the performance of actions with imaginary objects, such as food or, in one of the most detailed and well-known examples, a pull-toy.

The use of dolls for common actions and routines (or the partial, nonfunctional imitation of fragments of serious routines, such as cleaning or brushing with the appropriate implements) corresponds to some early and simple manifestations of imaginative play in humans, and seems to fit well as continuous with the few rare examples recorded in wild and captive primates. Symbolic training and routine interaction with humans may be enhancing what spontaneously remains very rare and rudimentary. Indeed, some apes with symbolic training, such as the gorilla Koko, are reported to engage in pretend drinking and to add "slurping sounds" to the action of taking a cup to their lips, which would enhance the imaginary component through a process of appropriate contextualization.

The performance of actions with imaginary objects (reported by Hayes, 1951, in one chimpanzee and by Savage-Rumbaugh et al., 1998, in one chimpanzee and one bonobo) is a more complex type of imaginative play that may not appear spontaneously in children until 2.5 or 3 years of age (Lillard, 2002). Its appearance in the chimpanzee Viki at 16 months in such an elaborate format (including imaginary accidents, such as the toy getting stuck and the request of help to solve the imaginary problem, as described by Hayes) is all the more surprising—as surprising as the complete absence of any other instances of imaginary object play and the complete disappearance of this one from Viki's repertoire after only a few weeks.

Savage-Rumbaugh and colleagues' descriptions of chimpanzee and bonobo play with imaginary food are less systematic than Hayes's. For example, we do not know exactly how the games first appeared: Were they completely spontaneous inventions, as were Viki's, or were they induced by the human caretakers? Inducement by the caretakers would not necessarily diminish the significance of the finding, as many instances of symbolic play in human children occur in the context of such an inducement and support. An interesting point is that these games appear to start from the imaginary snatching of a piece of fruit depicted on a photograph. The photograph may have provided the initial representational support for the production of the behavior, which later achieved higher levels of elaboration. These are, in our view, the more convincing and better documented examples of imaginative play in symbolically trained apes. Why should these apes manifest more frequent and more complex instances of imagination in play than their wild or symbolically naive peers?

Apes in the House

Linguistically trained apes show superior performance on a number of tests, including complex discrimination of relationships and imitation and joint attention (Gómez, 2004). Two approaches try to account for this finding. Premack (1983) proposes that linguistic apes have learned a new way of representing the world—a more abstract representational code that allows them to tackle problems in a new way that may enhance their performance. Linguistic apes have an "up-graded" mind. In contrast, Tomasello (1999) proposes that it is the process of interaction with humans, rather than the putative learning of artificial symbols, that changes the apes' minds. Humans would shape their attentional and action patterns in a characteristic way that somehow may result in higher cognitive processes, à la Vygotsky (1930). In fact, from a Vygotskian point of view, both accounts—the role of symbols in controlling behavior, and the effect of human socialization—may be complementary rather than alternative (Gómez, 2004).

This combination of symbolic tuition and human patterns of interaction may be especially relevant for the case of imaginative play. The teaching of symbolic action (or a close substitute of it) and the encouragement and promotion of symbolic games may explain well the apparently superior imaginative play of trained apes. After all, symbolic play blossoms in human infants at the time when they are producing their first words and simple utterances and with the encouragement and support of adults and some objects especially designed for symbolic play—toys such as dolls. But for the rearing and training conditions to have an effect, the apes themselves must contribute something.

Apes have shown some spontaneous ability to produce action schemas out of context, including some possible examples of simple object substitution (Gómez & Martín-Andrade, 2002). Although with all likelihood these behaviors occur without any pretending or symbolic intent (they need not be accompanied by the imagined representations of the usual contexts and objects), they may provide a basis on which the symbolic training builds to produce more complex instances of playful actions.

Zones of Proximal Evolution

The relative rarity and limitations of the imaginative play produced by even linguistically trained apes may make us conclude that these observations, curious as they are, may tell us little about the exuberant and apparently all-encompassing playful imagination of human children. However, as suggested earlier, the early manifestations of symbolic play in humans may not involve real fantasy and, in fact, may present several parallelisms with the instances observed in apes.

One suggestion is that symbolic and imaginative play in humans emerges in parallel with, and as a reflection of the development of, a number of specific adaptations for symbolization: imitation, nonverbal referential communication, labeling, and elaborate object manipulation and categorization. Symbolic play may be a mechanism that favors and facilitates these developmental lines, or it may be a mere by-product of them (see Smith, Chapter 8, this volume). Apes, however, lack several of these specific adaptations: They are not hard-wired for symbolic development, they are not primed for interpersonal imitation, their communication abilities only partially coincide with those of human infants, and the way in which they represent objects, although similar in their basic building blocks, may subtly differ from that of humans (Gómez, 2004). The cognitive and motivational makeup of apes is neither designed for symbolic play nor allows for a ready-made manifestation of it as a collateral derivative of other skills. However, when placed in an environment that offers new pressures and opportunities (human hand rearing and a regime of symbolic training), the cognitive makeup of apes appears to be able to generate some instances of the activities that we call imaginative play. They may be clumsy and relatively infrequent, because they may not be produced with exactly the same articulation of mechanisms as is human pretense, but they may reflect some of the genuine components of which fantasy is made: for example, (1) some ability to separate action schemas from the contexts in which they were acquired, even from the typical objects upon which they are executed; (2) some ability to recognize similarities between objects and physical representations of real objects such as dolls; and (3) an ability and a motivation to play with and explore action schemas.

In this sense, the apes' minds, when placed in a typical human environment, demonstrate some of the components from which fantasy may have emerged in human evolution. The human rearing and training given to the apes may create a "zone of proximal evolution"—a sketch of the direction in which an ape mind might begin to change if subject to certain pressures (Gómez, 2004; Gómez & Martín-Andrade, 2002). Of course, actual evolution could only occur if chance changes in the cognitive hardware of the brain were systematically selected. But a zone of proximal evolution may give us an idea of what kinds of minds are better fitted to begin a particular evolutionary race—in this case, the invention of symbols.

In the case of imaginative play, we are confronted with the additional complication that we might be dealing, not with the selection of a specific mechanism for symbolic play, but with the selection of a range of mechanisms with different functions, one of which—maybe not a primary one—is symbolic play. It is probably no coincidence that the few nonhuman examples of possible figurative drawing have been reported in some "linguistic" apes (although these appear to be even rarer than symbolic play; see, e.g., Patterson, 1978). Nonsymbolically trained apes may show great interest in scribbling and painting, and their pictorial activity may develop along lines roughly similar to those of young human infants, but they stop short of producing representational drawings (Morris, 1962).

SUMMARY

The issue of fantasy and symbolic play in apes is difficult to evaluate because we do not possess an appropriate database of observations (let alone, experiments) on which to reach conclusions. Very few studies have specifically addressed this issue, and, as a result, most reports are incidental (when not merely anecdotal) and lacking in crucial detail. Any conclusions are therefore tentative. Our own tentative conclusions are that symbolic play is not a typical behavior pattern of apes, in contrast to humans. Spontaneously, only some precursors or very rudimentary manifestations are very occasionally seen in wild and captive apes. Symbolic training and so-called "enculturation" have the effect of producing more frequent and more complex manifestations of imaginative play, some of which may involve actual imagination and pretense (although more detailed and longitudinal descriptions are needed to decide this issue). We propose that, even adopting a rich interpretational stance for the apes and a skeptical stance for the early manifestations of human children, there is still evidence of an important difference in quantity and quality of imaginative play between apes and humans. Our suggestion is that this difference reflects the evolution of specific adaptations for explicit symbolization and representation, communication and imitation, in humans. Only some components of these cognitive processes exist in apes. However, the exceptional conditions of hand-reared and linguistically trained apes may be producing a genuine "laboratory culture" of the sort of cognitive interactions out of which human fantasy and imagination were made.

ACKNOWLEDGMENTS

This chapter was written with the support of Grant No. BSO2002-00161 from the Dirección General de Investigación Científica y Técnica.

REFERENCES

Bateson, G. (1955). A theory of play and phantasy. *Psychiatric Research Reports, 2*, 39–51.

Breuggeman, J. A. (1973). Parental care in a group of free-ranging rhesus monkeys (*macaca mulatta*). *Folia Primatologica, 20*, 178–210.

Byrne, R., & Whiten, A. (Eds.). (1988). *Machiavellian intelligence: social expertise and the evolution of intellect in monkeys, apes and humans.* Oxford, UK: Oxford University Press.

deWaal, F. (1982). *Chimpanzee politics: Power and sex among apes.* London: Cape.

deWaal, F. (1986). Deception in the natural communication of chimpanzees. In R. W. Mitchell & N. S. Thompson (Eds.), *Deception: Perspectives on human and nonhuman deceit* (pp. 221–244). Albany: State University of New York Press.

Dunn, J., & Wooding, C. (1977). Play and its implications for learning. In B. Tizard & D. Harvey (Eds.), *Biology of play* (pp. 45–73). Philadelphia: Lippincott.

Fagen, R. (1981). *Animal play behavior.* New York: Oxford Univesity Press.

Fein, G., & Apfel, N. (1979). The development of play: Style, structure, and situation. *Genetic Psychology Monographs, 99*, 231–250.

Gardner, B. T., & Gardner, R. A. (1985). Signs of intelligence in cross-fostered chimpanzees. *Philosophical Transactions of the Royal Society of London (B), 308*, 159–176.

Gardner, R. A., & Gardner, B. T. (1969). Teaching sign language to a chimpanzee. *Science, 165*, 664–672.

Gardner, R. A., Gardner, B. T., & Van Cantfort, T. E. (1989). *Teaching sign language to chimpanzees.* New York: State University of New York Press.

Gómez, J. C. (2004). *Apes, monkeys, children and the growth of mind.* Cambridge, MA: Harvard University Press.

Gómez, J. C., & Martín-Andrade, B. (2002). Possible precursors of pretend play in non-pretend actions of captive gorillas (*Gorilla gorilla*). In R. W. Mitchell (Ed.), *Pretending and imagination in animals and children* (pp. 255–268). Cambridge, UK: Cambridge University Press.

Goodall, J. (1986). *The chimpanzees of Gombe.* Cambridge, MA: Harvard University Press.

Hayaki, H. (1985). Social play of juvenile and adolescent chimpanzees in the Mahale Mountain National Park, Tanzania. *Primates, 26*, 343–360.

Hayes, C. H. (1951). *The ape in our house.* New York: Harper & Row.

Hayes, K. J., & Hayes, C. (1952). Imitation in a home-raised chimpanzee. *Journal of Comparative and Physiological Psychology, 45*, 450–459.

Huttenlocher, J., & Higgins, E. T. (1978). Issues in the study of symbolic play development. *Minnesota Symposia on Child Psychology, 11*, 98–140.

Inhelder, B., Lézine, I., Sinclair, H., & Stambak, M. (1972). *Les débuts de la fonction symbolique* [The beginnings of the symbolic function]. *Archives de Psychologie, 163*, 189–293.

Jarrold, C., Carruthers, P., Smith, P. K., & Boucher, J. (1994). Pretend play: Is it metarepresentational? *Mind and Language, 9*, 445–468.

Jensvold, M. L. A., & Fouts, R. S. (1993). Imaginary play in chimpanzees (*Pan troglodytes*). *Human Evolution, 8*, 217–227.

Kellogg, W. N., & Kellogg, L. A. (1933). *The ape and the child.* New York: McGraw-Hill.

Ladygina-Kohts, N. N. (2002). *Infant, chimpanzee and human child.* Oxford, UK: Oxford University Press. (Translated from Russian, original work published 1935)

Leslie, A. (1987). Pretense and representation: The origins of "theory of mind." *Psychological Review, 94*, 412–426.

Lillard, A. (1994). Making sense of pretence. In C. Lewis & P. Mitchell (Eds.), *Children's early understanding of mind* (pp. 211–234). Mahwah, NJ: Erlbaum.

Lillard, A. (2002). Just through the looking glass: Children's understanding of pretence. In R. W. Mitchell (Ed.), *Pretending and imagination in animals and children* (pp. 102–114). Cambridge, UK: Cambridge University Press.

Matevia, M. L., Patterson, F., & Hillix, W. A. (2002). Pretend play in a signing gorilla. In R. W. Mitchell (Ed.), *Pretending and imagination in animals and children* (pp. 285–304). Cambridge, UK: Cambridge University Press.

Matsuzawa, T. (1999). Communication and tool use in chimpanzees: Cultural and social contexts. In M. D. Hauser & M. Konishi (Eds.), *The design of animal communication* (pp. 645–671). Cambridge, MA: MIT Press.

Mignault, C. (1983). *Transition entre les conduites sensori-motrices et l'activité symbolique chez le chimpanzé (Pan troglodytes)* [Transition between sensorimotor behaviors and symbolic activity in the chimpanzee]. Unpublished master's thesis, Departement de Psychologie, Université de Montréal, Montréal.

Mignault, C. (1985). Transition between sensorimotor and symbolic activities in nursery-reared chimpanzees (*Pan troglodytes*). *Journal of Human Evolution, 14*, 747–758.

Miles, H. L. (1991). The development of symbolic communication in apes and early hominids. In W. V. Raffler-Engel, J. Wind, & A. Jonker (Eds.), *Studies in language origins* (pp. 9–20). Amsterdam: Benjamins.

Miles, H. L., Mitchell, R. W., & Harper, S. E. (1996). Simon says: The development of imitation in an enculturated orangutan. In A. Russon, K. Bard, & S. T. Parker (Eds.), *Reaching into thought* (pp. 278–299). New York: Cambridge University Press.

Mitchell, R. W. (2002a). Imaginative animals, pretending children. In R. W. Mitchell (Ed.), *Pretending and imagination in animals and children* (pp. 3–22). Cambridge, UK: Cambridge University Press.

Mitchell, R. W. (Ed.). (2002b). *Pretending and imagination in animals and children.* Cambridge, UK: Cambridge University Press.

Morris, D. (1962). *The biology of art.* London: Methuen.

Parker, S. T., & McKinney, M. L. (1999). *Origins of intelligence.* Baltimore: Johns Hopkins University Press.

Patterson, F. (1978). Conversations with a gorilla. *National Geographic, 154*, 438–465.

Patterson, F. (1980). Innovative uses of language by a gorilla: A case study. In K. Nelson (Ed.), *Children's language* (Vol. 2, pp. 497–561). New York: Gardner Press.

Patterson, F., & Linden, E. (1981). *The education of Koko.* New York: Holt.

Perner, J. (1991). *Understanding the representational mind.* Cambridge, MA: MIT Press.

Perner, J., Baker, S., & Hutton, D. (1994). *Prelief*: The conceptual origins of belief and pretence. In C. Lewis & P. Mitchell (Eds.), *Children's early understanding of mind* (pp. 261–286). Mahwah, NJ: Erlbaum.

Piaget, J. (1945). *La formation du symbole chez l'enfant* [Play, dreams, and imitation in childhood]. Neuchâtel: Delachaux et Niestlé.

Power, T. G. (2000). *Play and exploration in children and animals.* Mahwah, NJ: Erlbaum.

Premack, D. (1976). *Intelligence in ape and man.* Hillsdale, NJ: Erlbaum.

Premack, D. (1983). The codes of man and beasts. *The Behavioral and Brain Sciences, 6*, 125–167.

Premack, D., & Premack, A. J. (1983). *The mind of an ape.* New York: Norton.

Rumbaugh, D. M. (Ed.). (1977). *Language learning in a chimpanzee: The LANA project.* New York: Academic Press.

Russon, A. (2002). Pretending in free-ranging rehabilitant orangutans. In R. W.

Mitchell (Ed.), *Pretending and imagination in animals and children* (pp. 229–240). Cambridge, UK: Cambridge University Press.

Russon, A., Vasey, P. L., & Gauthier, C. (2002). Seeing with the mind's eye: Eye-covering play in orangutans and Japanese macaques. In R. W. Mitchell (Ed.), *Pretending and imagination in animals and children* (pp. 241–254). Cambridge, UK: Cambridge University Press.

Savage-Rumbaugh, S., & Lewin, R. (1994). *Kanzi: The ape at the brink of the human mind.* London: Doubleday.

Savage-Rumbaugh, S., Shanker, S. G., & Taylor, T. J. (1998). *Apes, language, and the human mind.* Oxford, UK: Oxford University Press.

Smith, P. K. (1982). Does play matter? Functional and evolutionary aspects of animal and human play. *Behavioral and Brain Sciences, 5*, 139–155.

Smith, P. K. (2002). Pretend play, metarepresentation, and theory of mind. In R. W. Mitchell (Ed.), *Pretending and imagination in animals and children* (pp. 129–141). Cambridge, UK: Cambridge University Press.

Tomasello, M. (1999). *The cultural origins of human cognition.* Cambridge, MA: Harvard University Press.

Tomasello, M., Kruger, A. C., & Ratner, H. H. (1993). Cultural learning. *Behavioral and Brain Sciences, 16*, 495–552.

van Lawick-Goodall, J. (1968). The behaviour of free-living chimpanzees in the Gombe Stream area. *Animal Behaviour Monographs, 1*, 161–311.

Vygotsky, L. (1930). Tool and sign in the development of the child. In R. W. Rieber (Ed.), *The collected works of L.S. Vygotsky. Vol. 6. Scientific legacy* (pp. 1–68). New York: Kluwer/Plenum.

Whiten, A., & Byrne, R. W. (1988). Tactical deception in primates. *Behavioral and Brain Sciences, 11*, 233–273.

Wrangham, R. W., & Peterson, D. (1996). *Demonic males.* Boston: Houghton Mifflin.

Zeller, A. (2002). Pretending in monkeys. In R. W. Mitchell (Ed.), *Pretending and imagination in animals and children* (pp. 183–195). Cambridge, MA: Cambridge University Press.

CHAPTER 8

Social and Pretend Play in Children

PETER K. SMITH

Most of our evidence on human play comes from modern Western or Western-influenced industrialized societies. As has often been pointed out, these societies differ greatly from the kind of environments in which humans evolved. Thus, the range of behaviors and their functional significance may differ from what was true thousands of years ago, and from what might apply even now to people who do not live in industrialized societies (although all people are now influenced, to some extent, by global culture).

Early works on human play, such as those by Spencer (1878/1898), Groos (1898, 1901), and Hall (1908) did in fact take a broad evolutionary and cultural perspective on it, albeit at a mainly descriptive level or, at least, without the benefit of more modern theorizing on evolution and play. This broad perspective did not continue. Although the distinguished anthropologist Margaret Mead did have a master's degree in child psychology, her writings on child rearing in different cultures (1928, 1930, 1935) emphasized the cultural relativism of her doctoral supervisor, Franz Boas. Indeed, the major trend of anthropological writing throughout the 20th century took such a perspective, deemphasizing commonalities and an evolutionary perspective. Furthermore, children's play was seldom treated as a major topic, with the notable exception of Helen Schwartzman's (1976, 1978) overview.

There was a major wave of interest in child development among psychologists and educators in the 1920s onward, especially in North America, where child psychology institutes and welfare stations were established in a number of cities. Observational methods were used to describe typical age and sex differences in children's behavior, especially children of preschool age. Descriptions of play naturally featured in these descriptions, although the perspective was primarily one of encouraging acceptable behavior and designing environments that fostered this outcome; a more detached perspective (in either cultural or species terms) was lacking (Smith & Connolly, 1972).

Observational methods fell out of favor among psychologists in the 1940s and 1950s, as a laboratory-based and behaviorist-influenced discipline took shape and sought to establish its "scientific" credentials. This trend dominated developmental psychology until the 1970s and 1980s, when ethologically influenced studies began to make an impact (Blurton Jones, 1972; Hutt & Hutt, 1970; McGrew, 1972). The last quarter of the 20th century saw some reinvigoration of studies of play, from a variety of perspectives, increased dialogue among psychologists, anthropologists, and biologists, and increased attempts at syntheses (e.g., Bekoff & Byers, 1998; Fagen, 1981; Power, 2000; Smith, 1982, 1984).

The study of play is clearly influenced by the general way in which play is viewed and valued. Although play appears to be valued, to some extent, in all cultures, this valuation may stem from different reasons and to different extents. In traditional societies, play among children can be seen as a way of relieving adults from immediate caregiving responsibilities (as in sibling caregiving, Weisner & Gallimore, 1977). In addition, the kinds of play that emulate work can be seen as practice for later work skills and roles (Bock, Chapter 10, this volume). In modern societies, and certainly modern industrial societies with formal schooling, play appears to have been "coopted" as a way of improving a child's cognitive, linguistic, and social development. Play is typically seen by educators as serving important functions in children's development.

Smith (1988) described how a "play ethos" developed from the 1920s through the 1980s—basically, an uncritical assumption about the long-term benefits of play. For example, Susan Isaacs, an influential researcher at the Institute of Education in London at this time, wrote in a book for nursery teachers that

> those who have watched the play of children have long looked upon it as Nature's means of individual education. Play is indeed the child's work, and the means whereby he grows and develops. Active play can be looked upon as a

sign of mental health; and its absence, either of some inborn defect, or of mental illness. (Isaacs, 1929, p. 9)

This view has remained influential among educators and policymakers. Another book for teachers and parents stated that "play is the very essence of life and the only means whereby the infant can learn anything. It remains the chief means of learning well into school years, certainly until reading has been completely mastered" (Tudor-Hart, 1955, p. 10). A Department of the Environment report in the United Kingdom (1973) stated that "the realisation that play is essential to development has slowly but surely permeated our educational system and cultural heritage" (p. 1). Psychology textbooks of the period often cited many functions of play, quite uncritically. More recently, a reader for childhood professionals stated that "if we value learning through play, then surely we need to teach through play. . . . Play makes a vital contribution to children's learning and development" (Napier & Sharkey, 2004, pp. 151–152).

Smith (1988) argued that this play ethos actually distorted the interpretations of many of the studies of object and pretend play conducted during the 1970s and 1980s, as play research started to gain momentum after the bleak period of the 1940s–1960s. This argument is considered later in this chapter. In fact, during the 20th century the play ethos also probably distorted the kinds of play studied. There were many studies of object and pretend play—the kinds of play seen by educators as beneficial and conducive to the growth of cognition and imagination. There were fewer psychological studies of exercise play (whose benefits were seen as limited to physical rather than psychological development), and very little on rough-and-tumble play until the advent of an ethological perspective (because play fighting is not seen as desirable by many teachers and educators: Pellegrini & Smith, 1998; Schafer & Smith, 1996).

STAGES IN THE DEVELOPMENT OF PLAY

Piaget (1951) described three main (though overlapping) stages in play, which he linked to stages of cognitive development. The first is sensorimotor play in infancy, appearing in the second half of the first year; here the infant plays with objects, making use of their properties (falling, making noises) to produce pleasurable effects—pleasurable for the infant, that is! The kinds of "secondary circular reactions" described by Piaget as part of this sensorimotor period appear to be playful in this way. Some time into the second year, as sensorimotor development nears completion, symbolic play (fantasy or pretend play) emerges; this stage lasts from around 15 months to 6 years. Finally, Piaget argued that games with rules (publicly agreed among the players) supercede symbolic play from around 6 years onward.

Later, and developing Piaget's ideas, Smilansky (1968) postulated four stages in the development of play, shown in Table 8.1. Smilansky's main contribution was to add a stage of constructive play with objects—play that is more focused and mature than the sensorimotor play of infancy but is not pretend play. It is certainly useful to include constructive play activities, but it is by no means certain that constructive play precedes dramatic play; rather, they seem to coexist during the preschool years following the sensorimotor period. It is also difficult to distinguish them: Interviews with children often reveal that what seemed to be just constructive play was actually fantasy—the tower of bricks may actually be a "space rocket" (Takhvar & Smith, 1990).

Rubin, Watson, and Jambor (1978) thought it useful to combine Parten's (1932) schema of social participation (solitary, parallel, associative, cooperative) with Smilansky's schema of play to create a play hierarchy. This nested category schema means that a preschool child's play can be categorized according to both schemas—for example, solitary practice play, or associative dramatic play. Categories higher up either scale were argued to be more mature. This schema is useful for coding children's activities, but it has two limitations. First, it implies a developmental progression that is not universally accepted; some solitary play can be quite mature, for example; and constructive play seems to coexist with dramatic play rather than precede it. Second, in common with much 20th-century human play research, it omits

TABLE 8.1. Smilansky's (1968) Four-Sequence Developmental Model of Play

Label	Description
Functional play	Simple body movements or actions with objects (e.g., banging bricks)
Constructive play	Making things with objects (e.g., building a tower with bricks)
Dramatic play	Acting out roles in a pretend game (e.g., pretending to be a doctor)
Games with rules	Playing in a game with publicly accepted rules (e.g., football, hopscotch)

important kinds of play, particularly, physical activity play, rough-and-tumble play, and language play (Pellegrini & Perlmutter, 1987; Takhvar & Smith, 1990).

SEX DIFFERENCES IN PLAY

Boys and girls obviously differ substantially in play and social behavior as early as the preschool years. Observations of 2- to 4-year-olds have found that boys tend to prefer playing with transportation toys and blocks and engaging in activities that involve gross motor movements, such as throwing or kicking balls or rough-and-tumble play; girls tend to prefer dolls, dressing up, and other forms of domestic play. Furthermore, the sex-role stereotype of the materials moderates boys' and girls' play. Boys' play with female-preferred toys, such as dolls, is less sophisticated than it is with male-preferred toys, such as blocks (Pellegrini & Perlmutter, 1989). Many activities, however, do not show a sex preference at this age; sex differences in behavior and sex segregation of activities increase notably in the primary school years through adolescence (Golombok & Fivush, 1994; Maccoby, 1998).

Sex differences in pretend play are observable during the preschool period. Some studies find that girls engage in pretend play both more frequently and at more sophisticated levels than do boys, but findings are inconsistent (Göncü, Patt, & Kouba, 2002) and seemingly dependent on the play environment, toys available, etc. Whereas the pretense of girls' play tends to involve domestic themes, the pretense of boys' play tends to be more physically vigorous, often co-occurring with play fighting and superhero themes (Fein, 1981; Holland, 2003; Pellegrini & Perlmutter, 1987; Smith, 1977). Furthermore, girls seem to use their more mature language abilities to advantage in play discourses (Göncü et al., 2002). Mothers tend to engage in more symbolic play with their daughters than their sons, and these mother–daughter interactions predict peer fantasy play (Bornstein, Haynes, O'Reilly, & Painter, 1996; Bornstein, Haynes, Pascual, Painter, & Galperin, 1999).

Even in nursery school, children tend to select same-sex partners for play, and by the time children are getting into team games at about 6 or 7 years old, sex segregation on the playground is much greater. Boys tend to prefer outdoor play and, later, team games, whereas girls prefer indoor, more sedentary activities, and often play in pairs. Boys more frequently engage in play fighting, and this form of play often is embedded in fantastic play themes (see Table 8.2). Girls remain more oriented toward adults (parents and teachers) longer into childhood. In a study of 10- to 11-year-old children on U.S. playgrounds, Lever (1978) found that boys more often played in larger mixed-age groups, whereas girls more often formed smaller groups of same-age pairs. Boys liked playing competitive team games that were more complex in their rules and role structure, and which seemed to emphasize "political" skills of cooperation, competition, and leadership in their social relations; girls put more emphasis on intimacy and exclusiveness in their friendships.

Explanations of these sex differences need to refer to both biological and cognitive-developmental factors. There is evidence of the effects of sex hormones on behavior: Collaer and Hines (1995, p. 92) stated that "the clearest evidence for hormonal influences on human behavioural development comes from studies on childhood play," referring especially to the relationship between masculinization of genetic females during fetal development and "tomboyism." However, biological factors do not explain the process of sex-role identification and variations in sex roles in different societies. The social learning theory approach postulates that parents and others reward or reinforce sex-appropriate behavior in children; for example, they may encourage nurturant behavior and doll play in girls, and discourage both in boys. Children may also observe the behavior of same-sex models and imitate them; for example, boys might observe and imitate the behavior of male figures in movies and television in their playful behavior.

Although reinforcement does seem to have some effect, a number of studies suggests that more complex explanations are needed (Fagot, 1985). In nursery schools, for example, teachers generally reinforce quiet, "feminine" behaviors and do not reinforce boys' play with transportation toys or, indeed, any kind of rough-and-tumble play. But this does not stop boys from engaging in these kinds of play. An alternative perspective focuses on the importance of sex-role identification and peer-group pressures. Boys seem to be influenced by other boys, and girls by other girls. The cognitive-developmental approach argues that the child's growing sense of gender identity—the awareness that one is a boy or a girl—is crucial. Children imitate same-sex models and follow sex-appropriate activities because they realize that this is what children of their own sex usually do. Maccoby and Jacklin (1974) called this behavior "self-socialization" because it does not depend directly on external reinforcement. Although both biology and adult reinforcement influence sex differences, the direct influence of the peer group seems to be a very powerful one (Maccoby, 1998).

This chapter focuses on social and pretend play, rather than on object play (see Pellegrini & Gustafson, Chapter 6, this volume) or rough-and-tumble play (see Fry, Chapter 4, this volume), to examine definitions and the forms of pretend play; consider arguments about its functional significance—whether such kinds of play have been evolutionarily selected for either immediate or deferred benefits to the playing individual; and whether or not play offers continuing or new benefits in modern urban environments.

PRETEND PLAY

Psychologists have conducted prolific studies of pretend play—sometimes called make-believe play or, by Piaget (1951), symbolic play. In pretend play, a child uses an object or action in a nonliteral way—that is, the object or action does not mean what it usually means (e.g., a chair "becomes" a bus, and the child's "beep-beep" vocalizations and rotary arm movements, "become" the bus horn and the driver moving the steering wheel). Pretend play involves make-believe actions, objects, and roles. Lillard (1994, p. 214) defines pretense as involving six defining features: a *pretender,* a *reality,* and a *mental representation* that is *projected onto reality,* with *awareness* and *intention* on the part of the pretender.

Piaget (1951) was one of the first psychologists to describe the beginnings of pretend play in detail, by recording the behavior of his children. He noted that at the age of 15 months, his daughter Jacqueline put her head down on a cloth, sucked her thumb, blinked her eyes—and laughed. He interpreted this sequence as Jacqueline pretending to go to sleep; indeed, she did similar actions with her toy animals a month or so later. Characteristically, young children engage in pretend actions first with themselves, then with a doll or teddy—a process called "decentration."

Many early studies of pretend play focused on observing children when they were put in a laboratory playroom (alone) with some objects. This method had the advantage of providing a standardized procedure, but its unnaturalness was a significant disadvantage. Basically, these were observations of solitary play. A more naturalistic study—an intensive longitudinal study of children playing at home—was reported by Haight and Miller (1993), who made videotapes of children engaging in pretend play, starting when they were 12 months old and continuing until they were 48 months old. The surprising finding they reported was that about 75% of pretend play was social—first with mothers or parents, later with friends (peers)—whereas only a minority of the pretend play they observed was solitary.

In terms of frequency, pretend play begins during the second year of life, peaks during the late preschool years, and declines during the primary school years (Fein, 1981); this time course is similar to that for object play and earlier than rough-and-tumble play. Based on their home observations,

Haight and Miller (1993) found that rates of pretend play were 0.06 minutes/hour for 12- to 14-month-old children (0.1% of time), increasing to 3.3 (5.5%) at 24 months, and 12.4 (21%) at 48 months. In preschools, pretend play has been reported to take up some 15% of the day (Field, 1994).

Pretend play increases in complexity as well as frequency through the preschool years. Based on both laboratory and more naturalistic studies, Howes and Matheson (1992) described stages in the development of social pretense with mothers and peers. It appears that the mother (or older partner, perhaps a sibling) has a "scaffolding" role of supporting the play a lot initially by, for example, suggesting and demonstrating actions. For instance, the mother might "give teddy a bath" and then hand the teddy to the infant. Thus a lot of early pretend play by the child is largely imitative; it tends to follow well-established "scripts" or story lines, such as "feeding the baby" or "nursing the patient." In addition, realistic props, such as miniature cups, help to sustain pretend play. As children reach 3 or 4 years of age, they are less reliant on older partners and realistic props. They take a more active role in initiating pretend play; they adapt less realistic objects (a block becomes a cup, for example) or even just imagine the object completely; and they show an awareness of play conventions and competently negotiate roles within play sequences.

Pretend play is influenced by social variables as well as play materials. Children who are securely attached to their mothers engage in more sophisticated pretend play with them, compared to those who are insecurely attached; they initiate more play interactions, and the play has a more positive emotional tenor (Bretherton, 1989; Roggman & Langlois, 1987). In pretend play with peers, children's pretense is more sustained and complex when they are playing with friends, compared to acquaintances (Howes, Droege, & Matheson, 1994). Probably, children feel more trust and safety with friends; plus, they have more shared knowledge and expectations to facilitate the negotiation of play scripts.

Smilansky (1968) argued that the play of children from "culturally deprived" backgrounds was impoverished in terms of content, duration and complexity. This contention led to a body of research that validated Smilansky's finding that children from lower socioeconomic backgrounds showed less fantasy play. However, these studies were criticized by McLoyd (1982) for poor methodology. Some failed to define social class adequately or confounded it with other variables such as race or school setting. It is likely that global statements about social class effects on play are unwarranted. Nevertheless, there are related proximal influences, such as nature of parent facilitation of play, materials available, material familiarity, and so on, that account for the relationships obtained.

Smilansky (1968) also popularized the idea of the importance of "sociodramatic play": pretend play involving social role playing in an extended story sequence, such as playing doctors and nurses and patients, or spacemen and aliens, or parents and children. Her ideas led to the key concept of "play tutoring" (considered later). Sociodramatic play is common in 3- to 6-year-olds. Two examples, involving English children (ages 3 and 4 years) observed in a playgroup, are given in Table 8.2. They illustrate a number of characteristics: pretend actions ("drinking," "locking a door," "swimming"), pretend vocalizations ("Hooo . . ."), use of substitute objects (chairs as a ship), use of imaginary objects (the pretend tea and cup in Charlotte's actions, the pretend arrow in Miles's tummy), and explicit role assignment ("You're Daddy, aren't you?", "I'm a captain"). The first example shows domestic, object-based pretend play that make use of what is seen at home—making tea, reading books. The second example, of boys' play, shows an integration of pretend elements with rough-and-tumble elements (here, mainly chasing); there is a story theme of sorts, but it shifts (fishes . . . sea captain . . . shark . . . sea devil). Unlike domestic themes, this kind of sociodramatic play is more fantastic, relying on stories and films such as *Jaws*. It was vigorous, active play, and indeed another boy watched slightly nervously, not joining in, throughout the engagement. Both examples exhibit children employing social skills in negotiation, some conflicts, and competitive vying for attention.

IMAGINARY OBJECTS AND IMAGINARY COMPANIONS

Pretend play often involves object substitution. A wooden block becomes a cake for a picnic, a stick or piece of lego becomes a gun for war play, a cardboard box becomes a boat. Thus, the presence of play props can facilitate play; for younger children (around 2–3 years) it helps if the real object bears some similarity to the pretend object. For older children, greater dissimilarity can be accommodated—a process called decontextualization. A further stage of this development is to *imagine* the object. Overton and Jackson (1973) showed that when asked to pretend to comb their hair or brush their teeth, most 3- and 4-year-olds used a body part—fingers, for example—as a substitute comb or toothbrush, whereas 6- to 8-year-olds usually imagined a comb or brush being held in their hand.

A particularly intriguing type of imaginary object is the *imaginary companion* who follows the child around, or needs to be fed at mealtimes, or gets tucked into bed with the child. Some one-quarter to one-half of chil-

TABLE 8.2. Examples of Sociodramatic Play (a) with Two Girls and a Boy and (b) with Four Boys

(a)

Helen is sitting in a playhouse that has a plastic cup and teapot on a table. Charlotte is outside with a pram and a teddy in it.

Helen: (*to Charlotte*) I'm just getting the tea ready! Come on, it's dinnertime now!

Charlotte: Wait! (*She wraps up the teddy in a cloth, in the pram; then she comes into the playhouse and sits opposite Helen at the table.*)

Helen: I made it on my own! I want a drink! (*She picks up the teapot.*) There's only one cup—for me! (*She pretends to pour from the teapot into the cup. Charlotte pretends to pour from the teapot into an imaginary cup, then pretends to drink.*) I'm having chips for my tea! No, we're not having fish, we're not having any fish, only chips! (*She pretends to eat chips.*) We're going to sit on the floor now, sit down here!

Helen and Charlotte sit on the floor with some books.

Charlotte: Shall we lock the door now? (*She pretends to lock and bolt the door of the play house.*)

Darren comes up toward the playhouse. Charlotte pretends to lock the door again.

Helen: (*to Charlotte*) No, he's Daddy. (*to Darren*) You're Daddy, aren't you?

Darren pushes at the playhouse door. Charlotte pretends to unbolt and unlock it.

Darren comes in and sits next to them. He accidentally pushes the door open again.

Helen: Ooohh. (*She pulls the door closed and pretends to lock it.*)

They all sit and look at books.

(b)

Miles, Jim, Gareth, and Anthony are playing together. (A fifth child, Simon, watches throughout, sucking his thumb.)

Miles: (*to Jim*) Get in the ship! Get in the ship! (*to Anthony, Gareth, and Jim*) On the ship, you lot! (*He runs off to one of a row of red chairs at the end of the room.*) On the shi-i-ip!

Gareth and Jim sit on chairs next to him. Anthony lies on the floor.

Gareth: Ahoy there!

Gareth gets off his chair and pretends to swim, running around making swimming arm movements.

Miles gets off his chair and lies on the ground with Anthony.

Jim: Two fish in the water! Fishes! (*He pushes Miles and Anthony.*)

Miles: I'm a dead captain! I've got an arrow in there. (*He pulls up his vest and pokes his tummy.*)

Jim: He's dead, he's dead!

Gareth: Good!

Gareth picks up Miles's legs and Jim picks up his arms. They try to lift him.

Jim: Carry him!

(continued)

TABLE 8.2. *(continued)*

Anthony comes to help them.

Jim: He's dead!

The three boys cannot lift Miles (who is quite heavy) and lay him down again.

Jim: Let's take the arrow out. (*He pretends to pull an arrow out of Miles's tummy.*)

Anthony: Leave him, leave him.

Jim: Let's leave him. Till he gets better.

The three boys go and sit back on the chairs.

Jim: Hooo . . . [ship's siren noise]

Miles: I'm a shark! (*He gets up and runs toward the others.*)

Anthony: Yikes!

Jim: Shark!

Miles chases Gareth, Anthony, and Jim around the room.

Gareth: He's coming!

Gareth, Anthony, and Jim run back toward the chairs and sit down; Anthony pretends to steer the "ship."

Anthony: Rrr-rrr.

Miles has tripped over at the end of his chase and bumped his knee. He gets up slowly.

Miles: A shark cut my leg open! (*He shows his knee to Anthony. Jim feels left out.*)

Jim: I'm a sea-devil!

Anthony: I'm a sea-devil!

Gareth: I'm a captain. (*He picks up a captain's cap.*)

Miles: No, can I be the captain now?

Miles and Gareth tug over the captain's cap.

Gareth: No, I'm the captain, I got it first. (*He keeps the cap and puts it on, sitting on his chair.*)

Miles lies down on the floor again.

Anthony: No, not again, again, again!

dren have some form of imaginary companion, especially between 3 and 8 years of age; they are mostly abandoned by age 10. Children with imaginary companions tend to engage in a lot of sophisticated pretend play, generally (Taylor, 1999; Taylor, Cartwright, & Carlson, 1993). Children are not confused about the imaginary status of imaginary companions; that is, they are aware that their imaginary companions are different from real friends.

WAR PLAY

Although children's pretend play tends to be seen positively by adults, and indeed is often modeled and encouraged, an exception is play with war toys. The term "war play" has been applied to games with toy guns, weapons, and combat figures, as well as pretend fighting or warfare. This kind of play has elicited concern from some writers, and it is banned in many preschools and kindergartens. Views of parents are divided. Costabile, Genta, Zucchini, Smith and Harker (1992) surveyed parents in Italy and England. Some actively discouraged war play: "I do not like it. In fact children who do war play become less sensible and less obedient" (p. 358). Some were uncertain or felt it should be allowed within limits: "Unsure. We don't encourage it, but don't discourage it, either" (p. 358). Some allowed it unconditionally: "A natural part of a child's development, just as they act out cooking, etc. War/fighting is featured in so many things it would be difficult and unnatural to exclude it from a child's life" (p. 358).

Researchers, too, are divided. Carlsson-Paige and Levin (1987, 1990) argue that war toys and combat figures encourage stereotyped, good-versus-evil aggressive scripts that impoverish children's imaginations and encourage actual aggressive behavior. However, they recognize the difficulties of attempting to ban such play entirely; children can make toy guns out of lego, if replica guns are banned in the nursery. As a solution they advocate that adults intervene to turn such play to more constructive, less aggressive, ends. Holland (2003) documents her shift from an original position of banning war toys, to a recognition that, for boys especially, rough-and-tumble-type fantasy activities are a common and enjoyable activity, and it can be counterproductive to ban them. Sutton-Smith (1988) argues that, for children, war play is clearly pretend and reflects an aspect of real life. As one boy said, when his father asked him not to use toy guns, "But Dad, I don't want to shoot anybody, I just want to play."

For most children such activities are natural, separated from real life, and probably do little, if any, harm. But it is easy—and probably wise—to feel uncomfortable when the activity becomes very prominent; and there is the possibility that sanctioning violent play can make matters worse for children who are already disturbed or have violent tendencies. Dunn and Hughes (2001) studied 40 "hard-to-manage" and 40 control children in London, filming them playing in a room with a friend when they were 4 years old. The "hard-to-manage" children showed more violent fantasy; the extent of violent fantasy (across both groups) was related to poorer language and play skills, more antisocial behavior, and less empathic understanding 2 years later at age 6. The developmental issues regarding war play have still to be resolved (Goldstein, 1995; Smith, 1994b).

CROSS-CULTURAL PERSPECTIVES OF PRETEND PLAY

Fully developed pretend play, including role play and sociodramatic play, seems universal in human societies, according to anthropological accounts. It is seen in hunter-gatherer societies (Eibl-Eibesfeldt, 1989; Gosso, Ota, Morais, Ribeiro, & Bussab, Chapter 9, this volume; Konner, 1976), where pretend play occurs in mixed-age peer groups; for example, children using sticks and pebbles to represent village huts or the herding of cows. Reviews of pretend play in non-Western societies by Schwartzman (1978) and Slaughter and Dombrowski (1989) mention over 40 articles describing pretend play. There are certainly variations in the amount and type of such play, and it can appear to be "impoverished" in some societies (Gaskins, 1999; Smilansky, 1968), but its presence, to some degree, appears to be ubiquitous.

Morelli, Rogoff, and Angelillo (2003) observed young children in four communities: the Efe of the Democratic Republic of the Congo, a traditionally hunter-gatherer people (now also doing some farming work); a Mayan agricultural town in San Pedro, Guatemala; and two middle-class European American communities—West Newton, Massachusetts, and Sugarhouse, Utah. Twelve children, ages around 2 to 3 years (range 27–46 months), were observed during waking hours in each community. The percentages of time children were observed in various work- and play-related activities are shown in Table 8.3.

The Efe and San Pedro communities are, in themselves, very different. The Efe people of the Ituri Forest are foragers who do some work for a neighboring farming community and have little experience of formal schooling (none for the sample families). The San Pedro Mayan people value schooling (to a limited degree) and work either at home (weaving, trading, carpentry) or as laborers or farmers. In the two middle-class U.S. communities, the parents had had a lot of formal schooling, and employ-

TABLE 8.3. Percentage of Time Spent by 2- to 3-Year-Old Children from Four Communities in Various Activities

	Efe (DR Congo)	San Pedro (Guatemala)	West Newton (Massachusetts)	Sugarhouse (Utah)
Access to work	73	52	30	29
Observing work	26	19	13	12
Emulation of work in play	12	15	4	3
Involved in lesson	0.6	0.4	4	3
In play with adult	4	3	17	16
In scholastic play with adult	0	0	4	5
Conversation with adult on child-related topic	1	2	17	12

ment of almost all the fathers and half the mothers was away from the home.

As Table 8.3 shows, the young children in the U.S. communities had much less possibility of observing adult work activities (production of goods and services, maintenance of household) than either the Efe or San Pedro children. They also less often actually observed an adult working. Efe and San Pedro children had much more opportunity to see adults working, and did so about twice as often. Actual instances of children working were rare in all communities at this young age (though not absent; some Efe children helped prepare food, and San Pedro children sold fruit, for example). But the opportunities for observing adult work had an impact on play: The Efe and San Pedro children emulated work in play much more often—for example, playing store, pretending to cut firewood, making tortillas out of dirt, pretending to shoot animals with a bow and arrow, comforting a doll.

By contrast, in both the U.S. communities, children were much more involved in play with an adult (usually the mother), an appreciable part of which was scholastic play (literacy- or numeracy-related-activities for fun, such as singing alphabet songs, reading a story)—absent in the lives of Efe and San Pedro children. Other kinds of play with adults included pretend, rough-and-tumble, and games—activities deliberately initiated, or engaged in, by adults with children. Finally, only in the U.S. communities were there appreciable frequencies of (1) conversations between children and adults on child-related topics (e.g., "Did you have a nice time playing on the swings?"); and (2) of lessons, or adult-based instruction, on any topic. In sum, the lesser opportunities for children to have access to, observe, and participate in work activities in modern societies appear to be compensated for by specialized adult involvement with children, often in play-related activities but with some scholastic and lesson-based aspects, which clearly gain in importance as children get older and participate in formal schooling.

Pretend play in non-Western foraging and pastoral societies is generally present but affected by social context (Göncü et al., 2002). Gaskins (1999) observed 1–5-year-old children in a Mayan village community in the Yucatan, Mexico. Pretend play was rare in these children. It was not encouraged by adults, and early work demands were placed on the children. Gosso, Morais, and Otta (2003) compared play in native Indian (Parakana), rural (seashore), and three urban groups in Brazil. All forms of play were observed in all groups. Native Indian children were lower than other groups on overall play but nevertheless exhibited quite a lot of pretend play as well as exercise play. Martini (1994) observed children's play in the Marquesas Islands in Polynesia; fantasy play constituted about 12% of all play episodes. Many involved simple, isolated episodes of using objects symbolically, such as making mud bananas; more complex episodes involved imitation of adult actions, such as fishing, hunting, preparing feasts, and followed standard scripts.

Pretend and sociodramatic play appear to be sensitive to adult involvement and encouragement. High levels of adult-initiated pretend and sociodramatic play (and other kinds of play, e.g., scholastic) are examples of parental investment (MacDonald, 1993). This investment can be seen in a positive light, as fostering skills that pretend and sociodramatic play might provide (see below). On the other hand, parents' interests are not identical with children's interests, and parents may be attempting to switch children from exercise and rough-and-tumble play (which they may find noisy and irritating) to more "educational" forms of play; this may or may not be in the child's own interests. Parents themselves may be manipulated by media, commercial, and manufacturing interests to purchase and "consume" toys, actions that are reinforced by a prevalent "play ethos" (Smith, 1994a; Sutton-Smith, 1986).

DOES PRETEND PLAY HAVE IMPORTANT FUNCTIONS IN DEVELOPMENT?

Is pretend play an important part of development? Was it selected for in evolution because it brought immediate or deferred benefits to the child that plays? And do contemporary children gain immediate or deferred developmental benefits from playing? This set of questions—especially, what such benefits might be—opens up a controversial issue that is far from being resolved.

There are certainly good reasons to suppose that pretend play should have some useful developmental benefits. First, it takes up appreciable time—as we saw, perhaps some 15% of free time in the preschool years. Second, it has a characteristic age progression, peaking in the 2–6-year age range and especially present in the preschool years of 3-, 4-, and 5-year-olds. Harris (1994, p. 256) argues that "the stable timing of its onset in different cultures strongly suggests a neuropsychological timetable and a biological basis." Third, it appears to be a cultural universal. Slaughter and Dombrowski (1989, p. 290) suggest that "children's social and pretend play appear to be biologically based, sustained as an evolutionary contribution to human psychological growth and development. Cultural factors regulate the amount and type of expression of these play forms." These arguments would imply the presence of some selection pressure(s) in our evolutionary history that facilitated pretend play—hence, implicating pretend play as one of the characteristics selected for in our evolutionary history (e.g., language, theory of mind, social negotiations and exchange) as having important benefits for individuals.

Furthermore, design features of pretend and sociodramatic play aid the emergence of various developmental competencies in the preschool years. Pretend play, at least when fully developed, involves some metarepresentational skills (i.e., an object is represented as something else in the mind), and it can be creative in the directions it takes. In sociodramatic play, for example, sophisticated language may be used, roles are negotiated, and information may be exchanged about real-life activities ("Doctors don't do that, silly!"). De Lorimier, Doyle, and Tessier (1995) found that children exhibited more intense involvement in the negotiations in pretense versus nonpretense contexts.

Opposing this view, it could be argued that there are marked individual and cultural variations in pretend play that (apart from clinical syndromes such as autism where there is little pretend play shown) do not obviously have strong correlations with, or effects on, the kinds of skills mentioned. Indeed, other forms of play, work activities, and educational activities can also provide practice in all these developmental competencies.

There is much relevant correlational evidence, but it has limited value in confirming or disconfirming functional hypotheses of this kind. Experimental evidence would be stronger, but, as we shall see, that commonly has had shortcomings and has not yet yielded decisive evidence.

MODELS OF PRETEND PLAY IN DEVELOPMENT

Smith (2002) suggested three possible models of the role of pretend play in development.

Model 1: Pretend play is a by-product of other aspect(s) of development, with no important developmental consequence(s) of its own.

Model 2: Pretend play facilitates developmental consequence(s) but is not essential if other expected developmental pathways are present.

Model 3: Pretend play is necessary for important developmental consequence(s); in the absence of pretend play, these developmental consequences will not occur or will be significantly held back.

A few caveats about the nature of these three models should be noted. First, they are models of the function of pretend play for individual development. There may be further complexities when we consider functions for others (e.g., parents) or the wider societal context. Second, through the course of human biocultural evolution there might have been shifts from one model to another (e.g., from model 3 to model 2 if cultural change provided more pathways to development, or from model 1 to model 2 if an existing ability could be put to new uses ("exaptation"). Third, models 2 and 3 contain issues about threshold effects: whether certain amounts, frequencies, or types of pretend play are necessary or sufficient to instigate developmental consequences.

Given such caveats, possible predictions regarding correlational and experimental evidence for each model are suggested in Table 8.4. Regarding correlational evidence, for example, all models would predict some positive correlations between measures of pretend play and measures of skill, as all improve as children get older. Crucial here is whether correlations are partialed for age (or an age-substitute such as IQ or language ability). If this is done, model 1 predicts zero correlations, model 2 predicts patchy correlations (the correlations would vary depending on what alternative developmental experiences are open to the children in the study), whereas model 3 would predict uniformly strong positive correlations. Similarly, the outcomes of experimental studies (which, in practice, have all been enhancement, rather than deprivation, studies) should vary clearly in relation to the model adopted.

TABLE 8.4. Predictions Regarding Correlational and Experimental Evidence for Each Model of the Function of Pretend Play in Development

	Correlational Evidence[a]	Experimental Evidence[b]
Model 1 (no important function)	Any correlations would be around zero, once age, general IQ, or language ability were eliminated out	No differences expected between groups if other experiences are equated
Model 2 (one of many routes)	Correlations would be positive but highly variable in size, depending on other circumstances	Any differences for pretend play groups would be dependent on the nature of the control group's experiences
Model 3 (essential for development)	Correlations would be consistently positive and of appreciable magnitude	Consistent differences favoring enhanced pretend play groups (or disadvantaging deprived play groups)

[a]Correlations of pretend play measures with other developmental skills.
[b]Comparing developmental outcomes for groups with enhanced or deprived pretend play experiences and control groups.

WHAT IS THE STATUS OF THE EVIDENCE?: RESEARCH IN THE 1970S AND 1980S

Consistent with the play ethos, a number of correlational and experimental studies were carried out in the 1970s and 1980s that linked pretend and sociodramatic play to a range of outcomes, including language skills, conservation abilities, creativity, and role-taking and perspective-taking abilities. The correlational studies generally produced positive findings, though not always partialing for age, and not always consistent. Reviews of this work (Christie & Johnsen, 1985; Smith, 1988), however, have focused more on a substantial body of experimental studies from this period.

Dansky (1980) suggested that pretend play facilitates associative fluency, and Golomb and colleagues (Golomb & Bonen, 1981; Golomb & Cornelius, 1977; Golomb, Gowing, & Friedman, 1982) argued that pretense assists conservation learning. These ideas were tested by experimental studies, including "play-training" studies. The results appeared to support these functions of play. The most substantial body of work, however, focused on sociodramatic play and stemmed from Smilansky's work (1968; Smilansky & Shefatya, 1990). Smilansky had argued that sociodramatic play was essential for normal development, and that if a child were deficient in it, intervention should be carried out in the preschool to encourage and enhance such play. She had observed that immigrant children in Israeli preschools did not show much sociodramatic play and were also behind in language and cognitive skills. Subsequently, other studies in the United States and elsewhere documented similar findings for children from disadvantaged backgrounds. These findings may reflect parental failure to encourage or "scaffold" pretend play at home, which results in an unstimulating environment for the children. Of course, it might be that an unstimulating environment leads to both poor sociodramatic play skills and poor linguistic and cognitive skills. Nevertheless, Smilansky and others hypothesized that sociodramatic play, in itself, contributes in an essential way to many aspects of development (model 3), because of the rich practice in language skills, social negotiation, and object transformation that such play entails.

As a result, a number of intervention studies were conducted to test the hypothesis. Smilansky, and many other investigators, found that it is quite possible to enhance children's sociodramatic play by having preschool teachers and staff model such play, encourage it, take children on sensory-stimulating visits (e.g., to hospitals, zoos), and provide suitable props and equipment. This surge of play tutoring led to studies comparing equivalent groups of children (e.g., classes from the same school) either experiencing play training or acting as a control group to allow for effects of age and general preschool experience. Differences between pre- and posttest performances on various developmental tasks were compared; greater improvements in the play-tutored children were interpreted as strong evidence that sociodramatic play really is important.

Through the 1970s and 1980s, a number of studies of this kind got positive results—indeed, it seemed that whatever outcome tests the researchers used, play-tutored children improved more! Subsequently, it was argued (Smith, 1988) that these results—as well as those of Dansky and Golomb and colleagues—were confounded by three main flaws:

1. Selective interpretation of results: administering multiple tests but only highlighting significant findings or expected trends ($p < .1$); and giving methodological excuses for nonsignificant findings.
2. Effects of experimental bias: allowing experimenter effects to intrude by not taking precautions for blind testing or scoring.
3. Use of inappropriate control groups: comparing a pretend play-enhanced group with a control group that not only had no pretense enhancement but also less verbal stimulation or adult involvement generally.

When further experiments were run to control for these mistakes, the benefits of play tutoring were generally not found (Christie & Johnsen, 1985; Hutt, Tyler, Hutt, & Christopherson, 1989; Smith, Dalgleish, & Herzmark, 1981; Smith & Syddall, 1978; Smith & Whitney, 1987). Both play and control groups improved with time and age, but it seemed that general adult stimulation was important, rather than specifically the pretend play. As Hutt and colleagues (1989, p. 116) commented: "We would seriously question the importance placed upon fantasy play as an aid to cognitive development." This body of critical work certainly argues against model 3, but it is consistent with model 2: Sociodramatic (or pretend) play is one way to gain skills, but no more so, generally, than other ways of actively engaging with the social and physical environments.

These studies mainly examined the cognitive functions of pretense. What about social functions? Bretherton (1989) argued that pretend play might help children explore and master emotional difficulties (e.g., fear of the dark, family conflicts). But it is "securely attached" children who show more elaborate, socially flexible play and more benign resolutions of pretend conflicts. "Insecure-avoidant" children show more aggressive and fewer nurturant themes and often become obsessive in their play. Bretherton concluded that her proposed function of pretend play is "least open to those most in need of it" (1989, p. 399). Gordon (1993) studied children who had experienced emotional trauma. If pretend play helped deal with this trauma, he hypothesized, such children might be expected to show more pretend play organized around resolving emotional issues. Instead, Gordon reported that these children showed global inhibition of pretend play, less resolution of negative affective experience, and disorganized and perseverative pretend play activities of a simple kind. These findings suggest that pretend play may be diagnostic of children's emotional condition but are not evidence that pretend play functions to help children who really need to gain emotional mastery.

THE 1990s: PRETEND PLAY AND THEORY OF MIND

In the 1990s, a new generation of studies on the cognitive benefits of pretense linked it to theory of mind. Theory-of-mind abilities involve understanding (representing) the knowledge and beliefs of others. Possession of theory of mind is usually measured by the understanding that another person may hold a *false belief.* Two paradigmatic tasks here are the Sally-Anne or unexpected transfer task, and the Smarties or unexpected object task (Mitchell, 1997). Knowledge and beliefs are, themselves, "representations" of reality, so theory of mind involves a representation of a representation—a second order or "metarepresentation."

There are reasons to suppose that pretend play might be relevant to theory-of-mind acquisition (Smith, 2002). Both appear absent in nonhuman species, with possible exceptions of simple forms in the great apes (Gómez & Martín-Andrade, Chapter 7, this volume); both appear present in all normal humans and societies; both may be absent or severely held back in cases of autism and a few other clinical syndromes; and both have a characteristic developmental trend, with main features developing in the period of 2–6 years. In addition, pretend play does seem to exercise a child's understanding of a play partner's understanding of the play convention, agreement on pretense, and so on; and it does involve representing an object or activity as something that it is not—perhaps similar to metarepresentation?

PRETEND PLAY AND METAREPRESENTATION

Leslie (1987, 1988) argued that even simple pretend play (such as object substitution) is an early indicator of metarepresentational abilities. Suppose a child is holding a banana. Leslie (1987) termed the representation of this real object ("This is a banana") a "primary representation." Now the child puts the banana up to his mouth and says "Hello, Mummy." According to Leslie, the child is pretending that the banana is a telephone. When engaging in this kind of object substitution, there is a second representation of the banana in the child's mind ("This banana is a telephone") that is "decoupled" from reality—a second-order representation, or a metarepresentation (literally, a representation of the primary representation). This perspective suggested a leading role for pretend play, given that object substitution pretense emerge at around 18 months, compared to 3–4 years for theory-of-mind tasks.

This view, in turn, was criticized by Lillard (1993) and Jarrold, Carruthers, Smith, and Boucher (1994), who argued that individual pretense would be metarepresentational only if children have some awareness of their pretense, rather than just imitating actions seen earlier, or responding to the stimulus properties of the object. Even the capacity of very young children to engage in mutual pretense can be explained without recourse to metarepresentations, if they rely on scripts or possess generalized knowledge about behavior sequences and scenarios that allows them to join in a mutual pretend play episode. For example, a child has seen an older child or adult pick up a banana and talk into it, and so does the same. A number of experimental studies has also provided strong evidence that children view pretense as mentalistic and subjective, or infer mental state in someone who is pretending, only by 3–5 years (Hickling, Wellman, & Gottfried, 1997; Joseph, 1998; Lillard, 1994; Rosen, Schwebel, & Singer, 1997). Harris (1991) has used a simulation approach to establish directly the non-metarepresentational nature of pretense. All a child need do is imagine what the world *could* be like. This imagining does not require metarepresentations, only the ability to simulate counterfactual reality by setting aside what one knows about the world. After a series of experimental studies, Harris and Kavanaugh (1993, p. 76) concluded that "children might be able to engage in pretence, including joint pretence, without diagnosing the mental state of their play partner or themselves."

These experimental studies are in line with the naturalistic studies that led to the summary made by Howes, Unger, and Matheson (1992) of developmental changes in pretense. At 25–30 months in peer–peer social pretend, "each partner's pretend reflects the same script but their actions show no within pair integration. Partners inform each other of the script by comments on their own pretend and telling the other how to act" (pp. 15–16). In other words, there is little sign of understanding any distinct mental state of the other. At 31–36 months, peer–peer play is elaborated to "joint pretend with enactment of complementary roles. Children discriminate between speech used for enactment and speech about enactment. Children assign roles and negotiate pretend themes and plans" (p. 16). By 37–48 months "children adopt relational roles, are willing to accept identity transformations and generate or accept instruction for appropriate role performance. Children negotiate scripts and dominant roles and use metacommunication to establish the play script and clarify role enactment" (p. 68). Metarepresentation in social pretend emerges from 31 to 36 months and is clearly evident by 37–48 months.

In summary, there are good reasons for rejecting Leslie's view that even simple object substitution pretend implies metarepresentation. Rather, this latter skill starts to be evident at around 3 or 4 years in more complex and social pretend play. This is also the age period in which first-order theory-of-mind abilities emerge. Is there evidence that complex pretend play has a role in facilitating theory of mind at this age? Some studies have linked attachment security to theory of mind, and presence of older siblings to theory of mind. For both sets of studies, social pretend play has been proposed as one candidate for the causal link.

Meins (1997) argued that sensitive mothers/caregivers show "mind-mindedness"; that is, they "treat their children as 'mental agents,' taking into account their comments, actions and perspective" (p. 108). In her sample, these mothers had children who succeeded better on theory-of-mind tasks at 4 years. Fonagy, Redfern, and Charman (1997) assessed 3- to 6-year-olds on the Separation Anxiety Test, a measure of attachment suitable for this age, and found that securely attached children scored better on a false-belief task, even when chronological age, verbal mental age, and a measure of social maturity were controlled. The reasons advanced by Meins and by Fonagy and colleagues for this link include quality of pretend play, quality of conversational exchange and elaboration, and competence in peer-group interaction.

There is an analagous set of findings concerning effects of older siblings. Ruffman, Perner, Naito, Parkin, and Clements (1998) used data from four studies and consistently found benefits for theory-of-mind development of having older siblings. These authors suggested numerous mechanisms for this link, including familiarity and intimacy in interactions, cooperative activity and pretend play, advanced pretend play (role assignment and enactment), shared jokes, spontaneous discussion of feelings, talk about causality and inner states, reasoning about moral issues, familiarity with deception, and management of conflict. Indeed, Perner, Ruffman, and Leekam (1994) suggest that "pretend play is perhaps our best candidate for a cooperative activity which furthers the eventual understanding of false belief" (p. 1236).

Social pretend play could foster theory-of-mind abilities through increased use of mental state terms. Howe, Petrakos, and Rinaldi (1998) examined play in sibling dyads of ages 5–6 years and found that dyads who frequently engaged in pretend were more likely to use internal-state terms, especially in high-level negotiations about play. Brown, Donelan-McCall, and Dunn (1996) observed interactions between 4-year-olds and their older siblings, a best friend, or their mother. Mental-state terms were used a lot with siblings and best friends, and the usage often occurred during pretend play. Mothers also used a lot of mental-state terms with their children, but the children did not use so many mental state terms with mother as with sibling or best friend. The researchers also found a significant correlation between children's mental state term use and their scores on a false-belief task.

These studies make a plausible case for some link between pretend play and theory of mind, but they do not really distinguish the three models described above. For example, skill in language development might be an underlying factor that explains both ability for sophisticated pretend play and theory-of-mind skills. A longitudinal study by Astington and Jenkins (1999) suggested that language is fundamental to theory-of-mind development;

controlling for age and earlier abilities, earlier language ability predicted later theory of mind, but earlier theory of mind did not predict later language ability.

CORRELATIONAL EVIDENCE OF THE RELATIONSHIP BETWEEN PRETEND PLAY AND THEORY OF MIND

Children at a particular age vary in the amount to which they participate in pretend play. There is some slight evidence that girls engage in more social pretend play (see earlier discussion) and have better theory-of-mind performance (Charman, Ruffman, & Clements, 2002). Comparing individual children, some studies have reported positive correlations between measures of pretend play and theory of mind, whereas other studies have not. A summary of results from commonly cited studies in this area is given in Table 8.5. Most, but not all, partial out for age or equivalents (IQ, language).

The overall pattern of correlations is very patchy. No single correlation reaches the .5 level. The majority are nonsignificant (and we may suppose some bias toward significant results in reporting and publication). General frequency or amount of pretend or fantasy play is not a correlate of theory of mind, except for 4-year-olds in Taylor and Carlson's (1997) study—but if pretend play is important in developing theory of mind (rather than vice versa), we might expect the correlation to be higher in 3-year-olds, because that is the age when theory of mind (as assessed by false-belief tasks) starts to emerge. Two studies, one by Astingon and Jenkins (1995) and one by Nielsen and Dissanayake (2000),[1] do find positive correlations with explicit role assignment, but Youngblade and Dunn (1995) do not. Two studies, one by Suddendorf, Fletcher-Flinn, and Johnston (1999) and one by Nielsen and Dissanayake (2000), do find correlations with imaginary object pantomime but not with using a body part as object pantomime. Overall, the correlational findings appear to support model 2. If pretend play had a strong causal role in theory of mind (model 3), we would expect a stronger and more consistent pattern of positive findings.

[1] In another reference covering the same material (Nielsen & Dissanayake, 2001), slightly different correlations are reported. This difference is due to the use of Pearson correlations in this book version, whereas the later written journal article uses (preferred) Spearman correlations (M. Nielsen, personal communication, April 2004).

TABLE 8.5. Range of Correlations of Fantasy or Pretend Play Measures, with Theory-of-Mind Measures, in Several Recent Studies

Study	Participants: country, number, sex, age	Fantasy measures	Correlations with theory of mind
Astington & Jenkins (1995)	United States *n* = 30 (15b, 15g) 37–65 months	Amount of fantasy play	Four false-belief tasks .16
		Joint proposals	.50** (.49** if age and language ability controlled for)
		Role assignment	.41** (.37** if age and language ability controlled for)
Youngblade & Dunn (1995)	United States *n* = 50 (23b, 27g) Pretend measures at 33 months; theory-of-mind measures at 40 months	MLU as control	False-belief tasks; and aggregate of affective labeling and affective perspective taking
		Total participation	ns; .28* (.26, ns if MLU controlled for)
		Diversity of themes	ns; ns
		Role enactment	.30* (.31* if MLU controlled for); ns
		Role play (explicit)	ns; ns
Rosen et al. (1997)	United States *n* = 45 (19b, 26g) 11 × 3 years 25 × 4 years 9 × 5 years	No controls	Three tasks: appearance–reality, doll false belief, cartoon false belief
		Identify pretense	.26, .16, .08
		Infer pretender's mental state	.31, .19, .47**
Taylor & Carlson (1997)	United States *n* = 152 (75b, 77g) 57 × 3 years 95 × 4 years	Correlations controlled for verbal intelligence and age	Composite of four tasks: appearance–reality, false belief, representational change, restricted view
		General fantasy/ pretense measure	Overall .16* For 3-year-olds, ns For 4-year-olds, .27*
		Imaginary companions	.20*
		Impersonation	.22*

(continued)

TABLE 8.5. *(continued)*

Study	Participants: country, number, sex, age	Fantasy measures	Correlations with theory of mind
Taylor & Carlson (1997) *(cont.)*		Imaginary companions (parent)	.08
		Impersonation (parent)	.21**
		Imaginative play predisposition	.07
		Impersonation of animal (parent)	.17*
		Pretend actions (self-directed)	.26**
		Favorite play activity (parent)	.11
		Pretend actions (object-directed)	.13
Lillard (1999)	United States *n* = 48 (29b, 19g) 33–67 months	PPVT control	Two false-belief tasks
		Impersonation	.15
		Substitute objects	.01
		Pretend block play	.27
		Brain pretend questions	.14
		Moe task	.39** (ns with PPVT controlled for)
Schwebel et al. (1999, Study I)	United States *n* = 31 (18b, 13g) 16 × 3 years 9 × 4 years 6 × 5 years	Correlations controlled for age	Doll false belief; appearance–reality task
		Imaginativeness	.06; .46**
		Solitary pretend play	.12; .27
		Joint pretend play	.14; .36*
Schwebel et al. (1999, Study II)	United States *n* = 54 (25b, 29g) 3–5 years	Correlations controlled for age	Doll false belief; cartoon false belief; appearance–reality task
		Transformations	.17; .03; .29*
		Solitary pretend play	.17; -.22; .07
		Joint pretend play	.18; .03; .36**

(continued)

TABLE 8.5. *(continued)*

Study	Participants: country, number, sex, age	Fantasy measures	Correlations with theory of mind
Suddendorf et al. (1999)	New Zealand *n* = 44 (24b, 20g) 36–59 months		Composite of a false belief and an appearance–reality task
		Imaginary object pantomime	.42** (with age partialed out, .25**)
		Body part as object pantomime	ns
Nielsen & Dissanayake (2000)	Australia *n* = 40 (20b, 20g) reduced to *n* = 31 for statistical analyses 36–53 months	Correlations controlled for age	Composite of three false-belief tasks
		Object substitution	.35*
		Imaginary play	.09
		Attribution of anima	.23
		Role assignment	.35*
		Role play	.24
		Imaginary object pantomime	.35*
		Body part as object pantomime	.09

Note. b, boys; g, girls; MLU, mean length of outcome; PPVT, Peabody Picture Vocabulary Test; ns, not significant but no correlation stated.
*$p < .05$; **$p < .01$.

EXPERIMENTAL STUDIES

Two studies by Dias and Harris (1988, 1990) examined the effects of make-believe play on deductive reasoning. In their first series of studies (1988), 5- to 6-year-olds were given syllogisms (e.g., "All fishes live in trees/Tot is a fish/Does Tot live in water?"). In a verbal-group condition, children were told: "I am going to read you some little stories about things that will sound funny. But let's pretend that everything in the stories is true." In a play-group condition, they were told: "Let's pretend that I am in another planet. Everything in that planet is different. I'll tell you what's going on there." The children in the play-group condition scored better on the syllogisms at both ages, on correct responses and justifications. Their second series of studies (1990) had a similar design, but contrasted three paired sets of conditions: with/without imagery ("Make a picture in your head"); with/without a planet context ("in another planet"); and with/without make-believe intonation. Their clearest result was the poor performance of the without-imagery/without-planet/without-make-believe intonation group, with imagery appearing to be the strongest cue.

Dias and Harris argued that setting current reality aside and imagining a fictive alternative may be important in understanding false beliefs and thus in theory-of-mind development. However, their studies suffered from some of the limitations of the earlier studies during the 1970s and 1980s. In the 1988 study, there was no blind testing (though scoring was done blind). In the 1990 study, there was neither blind testing nor, apparently, blind scoring; there were also only six children in each group. Furthermore, in all studies, the "let's pretend" instruction is present in all conditions. Subsequently, Leevers and Harris (1999) have done further studies within this paradigm, leading them to reinterpret the earlier work. They now argue that it is not the fantasy or pretend component but simply *any* instruction that prompts a logical approach to the premises that helps children suceed at these syllogistic tasks.

The only direct training study on pretend play and theory of mind, to date, is by Dockett (1998; personal communication, 1999). In one Australian preschool center, she worked with two groups of children, mean age 50 months (n = 15, 18) who attended morning and afternoon sessions. All children were pre- and posttested on measures of shared pretense (from observations) and on theory-of-mind ability (in the performance of several standard tasks). One group of children received play training for 3 weeks, consisting of sociodramatic play focused around a theme of a pizza shop. The control group experienced the normal curriculum. Analyses showed that from an equal baseline, the play-training group demonstrated significant increases in frequency and complexity of group pretense, relative to the control group. This group also improved significantly more on the theory-of-mind tests, both at posttest and at follow-up 3 weeks later.

This study provides the best evidence yet of a causal link from pretend play to theory of mind, but a number of shortcomings should be noted (Dockett, personal communication, 1999). The groups were small and not specifically matched for relevant characteristics; the training period was short; and the testing was not done blind to condition. The latter may be the most serious reservation, given previous findings about experimenter effects in studies of play and problem solving (Smith, 1988; Smith & Whitney, 1987).

Sobel and Lillard (2001) carried out an experimental study (not a play-training study) with 24 children ages 4–5 years. Using a repeated measures design, the researchers compared performances on fantasy pretense and false-belief tasks with ordinary pretense and false-belief tasks. (This study followed previous work by the authors suggesting that children showed more mental understanding of pretense when it involved fantasy characters such as the Lion King). The pretense tasks involved two troll figures who were running either "just like the Lion King" (fantasy condition) or "just like a cat" (ordinary condition). Children were asked whether the troll was pretending to be that character (i.e., the Lion King or the cat). After completing four pretense tasks, the children were given a false-belief task that used either a monster doll (fantasy condition) or a nondescript doll (ordinary condition). Although the children understood pretense better in the fantasy condition, the fantasy/ordinary condition had no significant impact on false-belief performance. In addition, there was no significant association between success on the pretense tasks and passing the false-belief tasks.

BENEFITS OF PRETEND PLAY: SUMMARY

There are clearly arguments that pretend play should have some developmental benefits. The main arguments fall into two categories:

1. An evolutionary argument: Pretend play is universal across cultures, and like any other activity it incurs time and energy costs; therefore there should be benefits. While not prescriptive about what the benefits might be, they should be universally applicable. Furthermore, given that the costs of pretend play are relatively low, benefits might also be low or modest in nature.
2. An educational argument: Based almost entirely on studies in Western societies, the contention is made that pretend play is associated with advanced language, use of mental-state terms, and practice of theory-of-mind–type understandings (perhaps especially with more "fantasy'" figures).

As we have seen, the correlational and experimental evidence does not support any strong model of the causative influence of pretend play on other cognitive or social skills, including theory of mind. Outcomes are patchy. If Table 8.4 is any guide, model 2 seems supported more than model 3 (results are not consistent enough) or model 1 (there are some positive findings).

Perhaps we need to return now to recent human cultural evolution and the influence of social context. In modern industrialized societies, where most studies have been conducted, pretend and sociodramatic play are often encouraged and fostered by parents in the home as well as teachers in nursery schools to an extent that probably greatly exceeds what children experienced in traditional societies. Additionally, such play involves many more "fantastic" roles and themes than are found in traditional societies, in which pretend play normatively focuses on actual work roles and activities of older people in that society. Thus, any facilitating effects of pretend and sociodramatic play in development that are found in modern societies (even in model 2) might not be true of traditional societies. The nature of the roles played differ, the whole social and educational contexts of supporting and scaffolding play by adults differ, and the desired outcomes (adult skills) differ.

Insofar as it is possible to train children in theory of mind, role taking, and so on, are we (consciously or unconsciously) co-opting sociodramatic play as a means of enhancing these abilities? MacDonald (1993) recommended play scaffolding or training as a useful parental investment in this light—useful, that is, for parents and for society (and possibly for the children, too). Put another way, we might be creating cultural conditions in which pretend and sociodramatic play become convenient vehicles for facilitating certain skills we see as desirable to develop in children (argument 2 above). This is not the same as saying that the facilitation of these skills was any kind of original function (in an evolutionary sense) for our human ancestors (argument 1 above). A crucial issue for the latter—if it is not too late already—is to examine the nature and outcomes of pretend play in hunter-gatherer societies. Pretend play occurred (see Gosso et al., Chapter 9, this volume) but perhaps not to such an extent, and perhaps in ways that facilitated subsistence skills (such as pounding grain; see Bock, Chapter 10, this volume) rather than theory-of-mind, creative, or role-taking abilities. Even pretend play that facilitates subsistence skills may characterize settled farming communities more than hunter-gatherer societies.

Thus, it seems reasonable to argue that pretend play was originally a spin-off of our greater intelligence and language skills (model 1), and that only as we evolved culturally into settled farming communities, then urban industrialized societies, did we start to find pretend play to be one useful way to facilitate subsistence skills (in farming communities) and more "creative" and cognitive skills (in modern urban societies). Indeed, a "play ethos" in modern societies might be functional in raising levels of scaffolding for pretend play (and hence actual levels of pretend play in children) well above what they might be normally. Whether this heightened scaffolding actually has all the benefits intended and expected is still to be fully demonstrated. At least, children enjoy these activities and are usually willing partners in this culturally inspired endeavor.

REFERENCES

Astington, J. W., & Jenkins, J. M. (1995). Theory of mind development and social understanding. *Cognition and Emotion, 9*, 151–165.

Astington, J. W., & Jenkins, J. M. (1999). A longitudinal study of the relationship between language and theory of mind development. *Developmental Psychology, 35*, 1311–1320.

Bekoff, M., & Byers, J. (Eds.). (1998). *Animal play: Evolutionary, comparative, and ecological approaches.* New York: Cambridge University Press.

Blurton Jones, N. G. (Ed.). (1972). *Ethological studies of child behaviour.* London: Cambridge University Press.

Bornstein, M. H., Haynes, O. M., O'Reilly, A. W., & Painter, K. M. (1996). Solitary and collaborative pretense play in early childhood: Sources of individual variation in the development of representational competence. *Child Development, 67*, 2910–2929.

Bornstein, M., Haynes, O. M., Pascual, L., Painter, K. M., & Galperin, C. (1999). Play in two societies. *Child Development, 70*, 317–331.

Bretherton, I. (1989). Pretense: The form and function of make-believe play. *Developmental Review, 9*, 383–401.

Brown, J. R., Donelan-McCall, N., & Dunn, J. (1996). Why talk about mental states? The significance of children's conversation with friends, siblings and mothers. *Child Development, 67*, 836–849.

Carlsson-Paige, N., & Levin, D. E. (1987). *The war play dilemma: Balancing needs and values in the early childhood classroom.* New York: Teachers College, Columbia University.

Carlsson-Paige, N., & Levin, D. E. (1990). *Who's calling the shots?: How to respond effectively to children's fascination with war play and war toys.* Philadelphia: New Society.

Charman, T., Ruffman, T., & Clements, W. (2002). Is there a gender difference in false belief development? *Social Development, 11*, 1–10.

Christie, J. F., & Johnsen, E.P. (1985). Questioning the results of play training research. *Educational Psychologist, 20*, 7–11.

Collaer, M. L., & Hines, M. (1995). Human behavioral sex differences: a role for gonadal hormones during early development? *Psychological Bulletin, 118*, 55–107.

Costabile, A., Genta, M. L., Zucchini, E., Smith, P. K., & Harker, R. (1992). Attitudes of parents towards war play in young children. *Early Educational Development, 3*, 356–369.

Dansky, J. L. (1980). Make-believe: A mediator of the relationship between play and associative fluency. *Child Development, 51*, 576–579.

Department of the Environment. (1973). *Children at play.* London: Her Majesty's Stationery Office.

de Lorimier, S., Doyle, A.-B., & Tessier, O. (1995). Social coordination during pretend play: Comparisons with nonpretend play and effects on expressive content. *Merrill–Palmer Quarterly, 41*, 497–516.

Dias, M., & Harris, P. L. (1988). The effect of make-believe play on deductive reasoning. *British Journal of Developmental Psychology, 6*, 207–221.

Dias, M., & Harris, P. L. (1990). The influence of the imagination on reasoning by young children. *British Journal of Developmental Psychology, 8*, 305–318.

Dockett, S. (1998). Constructing understandings through play in the early years. *International Journal of Early Years Education, 6*, 105–116.

Dunn, J., & Hughes, C. (2001). "I got some swords and you're dead!": Violent fantasy, antisocial behavior, friendship, and moral sensibility in young children. *Child Development, 72*, 491–505.

Eibl-Eibesfeldt, I. (1989). *Human ethology.* New York: Aldine de Gruyter.

Fagen, R. M. (1981). *Animal play behavior.* New York: Oxford University Press.

Fagot, B. I. (1985). Beyond the reinforcement principle: Another step toward understanding sex role development. *Developmental Psychology, 21*, 1097–1104.

Fein, G. (1981). Pretend play: An integrative review. *Child Development, 52*, 1095–1118.

Field, T. (1994). Infant day care facilitates later social behaviour and school performance. In H. Goelman & E. Jacobs (Eds.), *Children's play in day care centers* (pp. 69–84). Albany: State University of New York Press.

Fonagy, P., Redfern, S., & Charman, T. (1997). The relationship between belief–desire reasoning and a projective measure of attachment security (SAT). *British Journal of Developmental Psychology, 15*, 51–61.

Gaskins, S. (1999). Children's lives in a Mayan village: A case of culturally constructed roles and activities. In A. Göncü (Ed.), *Children's engagement in the world: Sociocultural perspectives* (pp. 25–61). New York: Cambridge University Press.

Goldstein, J. H. (1995). Aggressive toy play. In A. D. Pellegrini (Ed.), *The future of play theory* (pp. 127–147). Albany: State University of New York Press.

Golomb, C., & Bonen, S. (1981). Playing games of make-believe: The effectiveness of symbolic play training with children who failed to benefit from early conservation training. *Genetic Psychology Monographs, 104*, 137–159.

Golomb, C., & Cornelius, C. B. (1977). Symbolic play and its cognitive significance. *Developmental Psychology, 13*, 246–252.

Golomb, C., Gowing, E. D., & Friedman, L. (1982). Play and cognition: Studies of pretense play and conservation of quantity. *Journal of Experimental Child Psychology, 33*, 257–279.

Golombok, S., & Fivush, R. (1994). *Gender development.* Cambridge, UK: Cambridge University Press.

Göncü, A., Patt, M. B., & Kouba, E. (2002). Understanding young children's pretend play in context. In P. K. Smith & C. H. Hart (Eds.), *Blackwell handbook of childhood social development* (pp. 418–437). Oxford, UK: Blackwell.

Gordon, D. E. (1993). The inhibition of pretend play and its implications for development. *Human Development, 36*, 215–234.

Gosso, Y., Morais, M. L. S., & Otta, E. (2003). *Pretend play of Brazilian children: A window into different cultural worlds.* Manuscript submitted for publication.

Groos, K. (1898). *The play of animals*. New York: Appleton.
Groos, K. (1901). *The play of man*. London: Heinemann.
Haight, W., & Miller, P. J. (1993). *Pretending at home: Early development in a sociocultural context*. Albany: State University of New York Press.
Hall, G. S. (1908). *Adolescence*. New York: Appleton.
Harris, P. L. (1991). The work of the imagination. In A. Whiten (Ed.), *Natural theories of mind* (pp. 283–304). Oxford, UK: Blackwell.
Harris, P. L. (1994). Understanding pretence. In C. Lewis & P. Mitchell (Eds.), *Children's early understanding of mind* (pp. 235–259). Hove: Erlbaum.
Harris, P. L., & Kavanaugh, R. D. (1993). Young children's understanding of pretense. *Monographs of the Society for Research in Child Development, 58*(1, Serial No. 231).
Hickling, A., Wellman, H. M., & Gottfried, G. M. (1997). Preschoolers' understanding of others' mental attitudes toward pretend happenings. *British Journal of Developmental Psychology, 15*, 339–354.
Holland, P. (2003). *We don't play with guns here*. Philadelphia: Open University Press.
Howe, N., Petrakos, H., & Rinaldi, C. M. (1998). "All the sheeps are dead. He murdered them.": Sibling pretense, negotiation, internal state language, and relationship quality. *Child Development, 69*, 182–191.
Howes, C., Droege, K., & Matheson, C. C. (1994). Play and communicative processes within long- and short-term friendship dyads. *Journal of Social and Personal Relationships, 11*, 401–410.
Howes, C., & Matheson, C. C. (1992). Sequences in the development of competent play with peers: Social and pretend play. *Developmental Psychology, 28*, 961–974.
Howes, C., Unger, O., & Matheson, C. C. (1992). *The collaborative construction of pretend*. Albany: State University of New York Press.
Hutt, S. J., & Hutt, C. (1970). *Direct observation and measurement of behavior*. Springfield, IL: Thomas.
Hutt, S. J., Tyler, S., Hutt, C., & Christopherson, H. (1989). *Play, exploration and learning: A natural history of the preschool*. London: Routledge.
Isaacs, S. (1929). *The nursery years*. London: Routledge & Kegan Paul.
Jarrold, C., Carruthers, P., Smith, P. K., & Boucher, J. (1994). Pretend play: Is it metarepresentational? *Mind and Language, 9*, 445–468.
Joseph, R. M. (1998). Intention and knowledge in preschoolers' conception of pretend. *Child Development, 69*, 966–980.
Konner, M. (1976). Relationships among infants and juveniles in comparative perspective. *Social Sciences Information, 13*, 371–402.
Lever, J. (1978). Sex differences in the complexity of children's play and games. *American Sociological Review, 43*, 471–483.
Leevers, H. J., & Harris, P. L. (1999). Persisting effects of instruction on young children's syllogistic reasoning with incongruent and abstract premises. *Thinking and Reasoning, 5*, 145–173.
Leslie, A. M. (1987). Pretence and representation: The origins of "theory of mind." *Psychological Review, 94*, 412–426.

Leslie, A. M. (1988). Some implications of pretence for mechanisms underlying the child's theory of mind. In J. W. Astington, P. L. Harris, & D. R. Olson (Eds.), *Developing theories of mind* (pp. 19–46). Cambridge, UK: Cambridge University Press.

Lever, J. (1978). Sex differences in the complexity of children's play and games. *American Sociological Review, 43*, 471–483.

Lillard, A. S. (1993). Pretend play skills and the child's theory of mind. *Child Development, 64*, 348–371.

Lillard, A. S. (1994). Making sense of pretence. In C. Lewis & P. Mitchell (Eds.), *Children's early understanding of mind* (pp. 211–234). Hove, UK: Erlbaum.

Lillard, A. S. (1999). Pretending, understanding pretense, and understanding minds. In S. Reifel (Ed.), *Play and culture studies. Vol 3: Theory in context and out* (pp. 233–254). Norwood, NJ: Ablex.

Maccoby, E. E. (1998). *The two sexes: Growing up apart, coming together.* Cambridge, MA: Belknap Press.

Maccoby, E. E., & Jacklin, C. N. (1974). *The psychology of sex differences.* Stanford, CA: Stanford University Press.

MacDonald, K. (1993). Parent–child play: An evolutionary perspective. In K. MacDonald (Ed.), *Parent–child play* (pp. 113–143). Albany: State University of New York Press.

Martini, M. (1994). Peer interactions in Polynesia: A view from the Marquesas. In J. L. Roopnarine, J. E. Johnson, & F. H. Hooper (Eds.), *Children's play in diverse cultures* (pp. 73–103). Albany: State University of New York Press.

McGrew, W. C. (1972). *An ethological study of children's behavior.* London: Academic Press.

McLoyd, V. C. (1982). Social class differences in sociodramatic play: A critical review. *Developmental Review, 2*, 1–30.

Mead, M. (1928). *Coming of age in Samoa.* New York: Morrow.

Mead, M. (1930). *Growing up in New Guinea.* New York: Morrow.

Mead, M. (1935). *Sex and temperament in three primitive societies.* New York: Morrow.

Meins, E. (1997). *Security of attachment and the social development of cognition.* Hove, UK: Psychology Press.

Mitchell, P. (1997). *Introduction to theory of mind: Children, autism and apes.* London: Arnold.

Morelli, G., Rogoff, B., & Angelillo, C. (2003). Cultural variation in children's access to work or involvement in specialized child-focused activities. *International Journal of Behavioral Development, 27*, 266–276.

Napier, N., & Sharkey, A. (2004). Play. In D. Wyse (Ed.), *Childhood studies: An introduction* (pp. 149–152). Oxford, UK: Blackwell.

Nielsen, M., & Dissanayake, C. (2000). An investigation of pretend play, mental state terms and false belief understanding: In search of a metarepresentational link. *British Journal of Developmental Psychology, 18*, 609–624.

Nielsen, M., & Dissanayake, C. (2001). A study of pretend play and false belief in preschool children: Is all pretense metarepresentional? In S. Reifel (Ed.), *The-*

ory in context and out: Play and culture studies (Vol. 3, pp. 199–215). Westport, CT: Ablex.

Overton, W. F., & Jackson, J. P. (1973). The representation of imagined objects in action sequences: A developmental study. *Child Development, 44*, 309–314.

Parten, M. B. (1934). Social participation among preschool children. *Journal of Abnormal and Social Psychology, 27*, 263–269.

Pellegrini, A. D., & Perlmutter, J. C. (1987). A re-examination of the Smilansky–Parten matrix of play behavior. *Journal of Research in Childhood Education, 2*, 89–96.

Pellegrini, A. D., & Perlmutter, J. C. (1989). Classroom contextual effects on children's play. *Developmental Psychology, 25*, 289–296.

Pellegrini, A. D., & Smith, P. K. (1998). Physical activity play: The nature and function of a neglected aspect of play. *Child Development, 69*, 577–598.

Perner, J., Ruffman, T., & Leekam, S. R. (1994). Theory of mind is contagious: You catch it from your sibs. *Child Development, 65*, 1228–1238.

Piaget, J. (1951). *Play, dreams and imitation in childhood.* London: Routledge & Kegan Paul.

Power, T. (2000). *Play and exploration in children and animals.* Mahwah, NJ: Erlbaum.

Roggman, L., & Langlois, J. (1987). Mothers, infants, and toys: Social play correlates of attachment. *Infant Behavior and Development, 10*, 233–237.

Rosen, C. S., Schwebel, D. C., & Singer, J. L. (1997). Preschoolers' attributions of mental states in pretense. *Child Development, 68*, 1133–1142.

Rubin, K. H., Watson, K. S., & Jambor, T. W. (1978). Free-play behaviors in preschool and kindergarten children. *Child Development, 49*, 534–536.

Ruffman, T., Perner, J., Naito, M., Parkin, L., & Clements, W. A. (1998). Older (but not younger) siblings facilitate false belief understanding. *Developmental Psychology, 34*, 161–174.

Schäfer, M., & Smith, P. K. (1996). Teachers' perceptions of play fighting and real fighting in primary school. *Educational Research, 38*, 173–181.

Schwartzman, H. B. (1976). The anthropological study of children's play. *Annual Review of Anthropology, 5*, 289–328.

Schwartzman, H. B. (1978). *Transformations: The anthropology of children's play.* New York: Plenum Press.

Schwebel, D. C., Rosen, C. S., & Singer, J. L. (1999). Preschoolers' pretend play and theory of mind: The role of jointly constructed pretence. *British Journal of Developmental Psychology, 17*, 333–348.

Slaughter, D., & Dombrowski, J. (1989). Cultural continuities and discontinuities: Impact on social and pretend play. In M. N. Block & A. D. Pellegrini (Eds.), *The ecological content of children's play* (pp. 282–310). Norwood, NJ: Ablex.

Smilansky, S. (1968). *The effects of sociodramatic play on disadvantaged preschool children.* New York: Wiley.

Smilansky, S., & Shefatya, L. (1990). *Facilitating play: A medium for promoting cognitive, socio-emotional and academic development in young children.* Gaithersburg, MD: Psychosocial and Educational Publications.

Smith, P. K. (1977). Social and fantasy play in young children. In B. Tizard & D. Harvey (Eds.), *Biology of play* (pp. 123–145). London: SIMP/Heinemann.

Smith, P. K. (1982). Does play matter? Functional and evolutionary aspects of animal and human play. *Behavioral and Brain Sciences, 5*, 139–155.

Smith, P. K. (1984). (Ed.). *Play in animals and humans.* Oxford, UK: Blackwell.

Smith, P. K. (1988). Children's play and it's role in early development: A re-evaluation of the "play ethos." In A. D. Pellegrini (Ed.), *Psychological bases for early education* (pp. 207–226). Chichester, UK: Wiley.

Smith, P. K. (1994a). Play training: An overview. In J. Hellendoorn, R. van der Kooij, & B. Sutton-Smith (Eds.), *Play and intervention* (pp. 185–194). Albany: State University of New York Press.

Smith, P. K. (1994b). The war play debate. In J. H. Goldstein (Ed.), *Toys, play and child development* (pp. 67–84). Cambridge, UK: Cambridge University Press.

Smith, P. K. (2002). Pretend play, metarepresentation and theory of mind. In R. W. Mitchell (Ed.), *Pretending and imagination in animals and children* (pp. 129–141). Cambridge, UK: Cambridge University Press.

Smith, P. K., & Connolly, K. (1972). Patterns of play and social interaction in pre-school children. In N. Blurton Jones (Ed.), *Ethological studies of child behaviour* (pp. 65–95). London: Cambridge University Press.

Smith, P. K., Dalgleish, M., & Herzmark, G. (1981). A comparison of the effects of fantasy play tutoring and skills tutoring in nursery classes. *International Journal of Behavioural Development, 4*, 421–441.

Smith, P. K., & Syddall, S. (1978). Play and nonplay tutoring in preschool children: Is it play or tutoring which matters? *British Journal of Educational Psychology, 48*, 315–325.

Smith, P. K., & Whitney, S. (1987). Play and associative fluency: Experimenter effects may be responsible for previous findings. *Developmental Psychology, 23*, 49–53.

Sobel, D. M., & Lillard, A. S. (2001). The impact of fantasy and action on young children's understanding of pretence. *British Journal of Developmental Psychology, 19*, 85–98.

Spencer, H. (1898). *The principles of psychology.* New York: Appleton. (Original work published 1878)

Suddendorf, T., Fletcher-Flinn, C., & Johnston, L. (1999). Pantomime and theory of mind. *Journal of Genetic Psychology, 160*, 31–45.

Sutton-Smith, B. (1986). *Toys as culture.* New York: Gardner Press.

Sutton-Smith, B. (1988). War toys and childhood aggression. *Play and Culture, 1*, 57–69.

Takhvar, M., & Smith, P. K. (1990). A review and critique of Smilansky's classification scheme and the "nested hierarchy" of play categories. *Journal of Research in Childhood Education, 4*, 112–122.

Taylor, M. (1999). *Imaginary companions and the children who create them.* New York: Oxford University Press.

Taylor, M., & Carlson, S. M. (1997). The relation between individual differences in fantasy and theory of mind. *Child Development, 68*, 436–455.

Taylor, M., Cartwright, B. S., & Carlson, S. M. (1993). A developmental investigation of children's imaginary companions. *Developmental Psychology, 29*, 276–285.

Tudor-Hart, B. (1955). *Toys, play and discipline in childhood.* London: Routledge & Kegan Paul.

Weisner, T. S., & Gallimore, R. (1977). My brother's keeper: Child and sibling caretaking. *Current Anthropology, 18*, 169–190.

Youngblade, L. M., & Dunn, J. (1995). Individual differences in young children's pretend play with mother and sibling: Links to relationships and understanding of other people's feelings and beliefs. *Child Development, 66*, 1472–1492.

PART V

HUNTER-GATHERERS AND PASTORAL PEOPLES

CHAPTER 9

Play in Hunter-Gatherer Society

YUMI GOSSO, EMMA OTTA, MARIA DE LIMA SALUM E MORAIS,
FERNANDO JOSÉ LEITE RIBEIRO, AND VERA SILVIA RAAD BUSSAB

The foraging way of life, also known as hunting-gathering, is regarded as the environment of human adaptedness, prevalent for more than 90% of the time *Homo sapiens* existed on earth. *Homo sapiens* appeared somewhere between 150,000 and 200,000 years ago. Long before, perhaps for 2 million years, the ancestors of modern humans had been leading a hunting-gathering way of life. The relatively recent advent of agriculture and animal husbandry could not have played a large role in shaping the modern human genotype. Ethologists and evolutionary psychologists consider that our basic psychological mechanisms have been shaped in the context of the hunting-gathering way of life to solve specific problems of adaptation (Bjorklund 1997; Bjorklund & Pellegrini, 2002; DeVore & Konner, 1970; Eibl-Eibesfeldt, 1989; Geary & Bjorklund, 2000; Smith, 1982).

HUNTER-GATHERER SOCIETIES AS A PRIVILEGED OPPORTUNITY OF GAINING INSIGHTS ABOUT HUMAN NATURE

The hunter-gatherer hypothesis means that small, permanent groups of individuals that comprised adult males, adult females, and their young were the typical arrangement within which human evolution took place. So humans in a hunter-gatherer group are in their natural physical, cultural, and psychological environment—so much so that our contemporary way of life can be seen as thwarting our nature in a manner similar to what occurs in other animals that fall short of manifesting their full potential when kept in artificial settings.

Modern developments have brought deep changes to the way of life of present-day humans; the benefits resulting from science and technology should not overshadow the fact that modern people are no longer living in their environment of adaptedness. In spite of increased life span and comfortable lifestyles, the large discrepancies that exist between ancestral and modern environments may create psychological problems and reduce the quality of life. Ethological and evolutionary psychological analyses suggest several ways in which modern psychological environments cause damage to humans (Buss, 2000). Typically, modern humans live in urban areas, within isolated nuclear families that often have no extended kinships, but surrounded by thousands of other humans. If psychological well-being depends on the existence of intimate connections with others, on being a valued member of an enduring social group, and being part of a network of extended kin, then the conditions of modern living may entail negative consequences and profound change.

Few human groups still maintain a predominantly foraging way of life; studying them may provide us with a unique opportunity to learn about human nature.

DATA SOURCES ON PLAY IN HUNTER-GATHERER SOCIETIES

Play is an intrinsically motivated activity; there is a universal psychological capacity to play that is part of the human condition, as is the capacity to speak (Sutton-Smith & Roberts, 1981). Although we do not yet know how important play is to the growing-up process, and its function remains a matter of debate (Pellegrini & Smith, 1998a; Smith, 1982), play seems to be serious business.

Field studies have greatly improved our knowledge about play. Students of play behavior (cf. Smith & Pellegrini, Chapter 11, this volume) have a special interest in field studies of play among children of hunter-gatherer societies, counterbalancing the predominance of research done with Western urban children. Our goal here is to review child play among extant foragers: Brazilian Indians (e.g., Parakanã, A'uwe-Xavante, Xocó, Guayaki, Kaingáng, Camaiurá, Mehináku, Tukano, and Canela) and Africans (e.g., !Kung, !Ko, Hazda, /Gwi). Unfortunately, many observations about child development and play in hunter-gatherer societies are anecdotal and have been reported as part of larger studies mainly interested in lifestyle and culture. Children appear in such reports because they happened to be present in the observation area. Few studies have focused specifically on child play (Bastos, 2001; Bichara, 1999, 2003; Gregor, 1982; Nunes, 1999, 2002).

Our major sources are studies of play in South American Indians, mainly Brazilian Indians (Figure 9.1) such as Parakanã (Gosso & Otta, 2003), A'uwe-Xavante (Nunes, 1999, 2002), Camaiurá (Moisés, 2003), and Mehináku (Gregor, 1982). As part of our data source we have also used studies on children's behavior in African hunter-gatherer societies (Blurton Jones, 1993; DeVore & Konner; 1970; Draper, 1976; Draper & Cashdan, 1988), maternal care (Hewlett, Shannon, Lamb, Leyendecker, & Schölmerich, 1998; Konner, 1977), and general way of life (Eibl-Eibesfeldt, 1974, 1989; Leakey, 1981; Shostak, 1976, 1981; Wannenburgh, 1979). Such studies refer mainly to the !Kung or the !Kung San, who live in Botswana and Namibia.

Present-day foragers are not exact equivalents of our ancestors. For example, the !Kung, who are the most studied African hunter-gatherers, live at the edge of the Kalahari desert, whereas in the past, they probably lived in a wealthier savannah environment. Likewise, the Parakanã Indians, whose children were directly observed by one of us (Y.G.) as a part of her doctoral thesis, do not prefectly replicate our ancestors in that they have a small-scale subsistence agriculture. However, hunting and gathering are routine activities responsible for supplying an important part of their diet. Moreover, they are unequivocally a tribal society; their culture is nearly intact in the sense that it strongly resembles pre-Columbian Brazilian Indians.

Yumi Gosso stayed at the Parakanã village four times between 1998 and 2003, totaling 9 months. During these periods, she conducted systematic observations of child behavior while also fully interacting with the Indians, using a mixture of her increasing knowledge of the Parakanã language and the rudiments of Portuguese known by some Parakanã individuals. Only some young men are reasonably fluent in Portuguese; old people and many women do not speak Portuguese at all, and the children speak only Parakanã.

Under governmental protection, present-day foragers may live in favorable environmental conditions, but their movement is limited, in contrast with the conditions in their ancestral environment. Figure 9.2 shows an aerial photograph of the Parakanã Indians Reserve at Pará, a state in northern Brazil. The reserve is surrounded by large areas of the forest, cleared of vegetation.

FIGURE 9.1. Geographic distribution of main Brazilian Indian groups mentioned in the text.

FIGURE 9.2. Aerial view of the Parakanã region. The reserve is the forested area on the right. Photograph by C. Emídio-Silva.

CHILDHOOD IN HUNTER-GATHERER SOCIETIES

In most hunter-gatherer societies, the needs of babies are readily satisfied; they are fed at request and usually never feel hungry. Infants are the center of attention; parents, older siblings, and other members of the community all participate in lulling the baby. During their first year of life, babies experience intense bodily contact with their mothers, who take them wherever they go, with the help of a sling. In this way the babies begin to get familiar with the world surrounding them (Gosso & Otta, 2003; Konner, 1972, 1977), as they ride along with their mothers to the field, during gathering, and during long walks through the forest.

When children reach the age of 2 or 3 years, there is a sharp change in their status, partly caused, typically, by the birth of a new baby. The child passes out of the "notorious" stage and becomes an "insignificant" member of the community. In Indian communities, it is common to see children 3 or 4 years old carrying small babies (Draper & Cashdan, 1988; Eibl-Eibesfeldt, 1989; Gosso & Otta, 2003; Lordelo & Carvalho, 1989; Mead, 1949; Nunes, 1999; Pereira, 1998).

In relation to childhood among the !Kung, Konner (1977) identifies three salient characteristics: (1) The mother–child relationship is bonded and long; (2) this relationship occurs in a dense social context, where there is contact with other adults; and (3) the closeness of the relationship with the mother is gradually replaced by interactions with mixed-age groups of children.

In Indian societies, children are usually free to move around the village with little or no restrictions. Hunter-gatherers do not give orders to their children; for example, no adult announces bedtime. At night, children remain around the adults until they feel tired and fall asleep (Draper, 1976; Gosso & Otta, 2003; Truswell & Hansen, 1976). In spite of their freedom, they learn about their social limits in subtle ways (Nunes, 1999).

Parakanã adults do not interfere with their children's lives. They never beat, scold, or behave aggressively with them, physically or verbally, nor do they offer praise or keep track of their development. Children do not go to parents for help or to complain about one another. In general, the parent–child relationship is much less intense, in both directions, than in non-foraging societies. This nuclear relationship is diluted in the midst of the large community. There are so many other children and adults around—always there—that the parent–child relationship is very different from the extremely dense psychological environment of families, nuclear or extended, in nontribal societies.

CLASSIFICATION OF PLAY BEHAVIOR

Our review of existing data on play in hunter-gatherer societies is organized according to a modified game typology based on Parker (1984):

1. Exercise play: wide locomotor movements (e.g., running, jumping, sliding, climbing).
2. Exercise play with objects: use of wide movements with objects (e.g., walking on stilts, flicking *zeni* into the air).
3. Object play: use of fine movements to move objects in some way (e.g., twisting tops, playing with bows and arrows).
4. Construction play: games in which materials are combined to create a new product (e.g., modeling sand, making baskets).

5. Social contingency play: games apparently motivated and reinforced by pleasure in producing contingent responses in others, and in responding contingently to others (e.g. peek-a-boo, tickling, imitating gestures or verbalizations).
6. Rough-and-tumble play: a social contingency game involving vigorous physical contact, wrestling, and pursuit-and-escape activities accompanied by playful signals (e.g., play face) that distinguish it from aggression.
7. Fantasy play: giving real or imaginary objects different properties from those they actually possess, creating and representing imaginary scenes (e.g., using a stick as a canoe, simulating domestic scenes, playing roles such as shaman).
8. Games with rules: games guided by explicit rules (e.g., field games with a ball, involving aimed throwing at long distances, and ground games involving dice throwing at short distances and turn taking).

EXERCISE PLAY

Forager children engage in vigorous and wide locomotion movements, mainly in the social context of play. Their own body is a source of play. Nunes (1999) described A'uwe-Xavante children running around the Namunkurá village not because they are in a hurry, but because their bodies "ask" for movement. They jump on one foot, alternating feet when they get tired; they walk backwards; they walk with their eyes closed or looking upward; they also turn around on their heels, with opened arms. Two children may take hands and spin together until they fall to the ground, dizzy and laughing heartily.

Similarly, Wannenburgh (1979) relates how !Kung children hook their right legs together, then hop round and round, clapping and singing, until one of them overbalances. This traditional game, called *n≠a n≠a hau,*[1] took its name from a species of thorn tree.

Trees can be a venue for different forms of exercise play. Kaingáng (Pereira, 1998), Mehináku (Gregor, 1982), and Parakanã (Gosso & Otta, 2003) Indian children were frequently seen climbing trees. Parakanã children, for example, were often observed climbing up and down guava and other trees, apparently with no specific purpose, just for the fun of it (Figure 9.3). While climbing, they would sometimes get a fruit, eat it, and continue with their activity. This type of play was usually observed among small groups of children on the same tree. A continuous climbing and descending activity on trees nearby the village was commonly seen in small children

FIGURE 9.3. Parakanã children playing in a tree. Photograph by Y. Gosso.

(3–5 years old). To get down from a tree, some children jumped or slid down it, controlling their movements by clutching the branches with their arms and hugging the tree trunk with their feet. Both boys and girls climbed trees, but only girls were seen helping others in this activity. Children were also frequently observed on high positions in the trees, looking down at what was happening on the ground (Gosso & Otta, 2003).

Handstands and human pyramids—in which one child climbs on another, trying to maintain balance—are also common activities among the Mehináku (Gregor, 1982) and the Camaiurá (Moisés, 2003). A variation of the pyramid play, called "manioc," was observed among the Camaiurá: A child lies on the ground and others lie on top, forming a pile, until the first child cannot keep his or her balance and lets the rest fall to the ground (Moisés, 2003).

Among the Mbuti Pygmy group of the Ituri forest in the Congo, Turnbull (1961) describes how children played a few yards from the main camp, in a place called *bopi*. There are always trees for the youngsters to climb. Sometimes six or more children would climb to the top of a young tree, bending it down until it touched the ground—and providing great entertainment. The fun is to synchronize the leap off the tree at once; if somebody is slower than the rest, he or she is hurled through the air as the tree comes back to the upright position, to the jeers and laughter of peers.

Most indigenous Brazilian villages are located near small rivers, essential for the people's lives. The river is used for bathing, fishing, and as an important transport route for covering great distances in canoes. The river is a very attractive place for the Indian children who live in a tropical climate. Very young babies are washed in the river, and they learn to hold their breath when bigger children or adults submerge them. It is common to see 3- or 4-year-old children riding on the backs of older girls as they swim and dive freely. Gradually, the very young children learn to swim. A'uwe-Xavante boys (Nunes, 1999, 2002) and Parakanã children (Gosso & Otta, 2003) freely jump into the river and swim, both following and against the flow.

Mehináku children from Mato Grosso (Gregor, 1982) and the Parakanã (Gosso & Otta, 2003) frequently use trees and stumps as diving boards to jump into the water. Even small children can be seen performing this type of activity without adult help or supervision. In Figure 9.4, a 4-year-old Parakanã boy jumps into the river of the Paranowaona village.

FIGURE 9.4. Parakanã children using a tree stump as a diving board. Photograph by Y. Gosso.

TABLE 9.1. Mean Percentage of Observation Time Spent in Various Types of Play by Parakanã Indians as a Function of Sex and Age Class[a]

Play typology	Descriptive statistics	*Konomia* Boys	*Konomia* Girls	*Otyaro* Boys	*Otyaro* Girls
Exercise	*M*	28.7	23.0	23.1	20.8
	SE	3.4	3.0	3.2	3.0
Construction	*M*	5.0	10.8	9.2	0.3
	SE	2.3	2.0	2.1	2.0
Social contingency	*M*	5.5	7.7	2.4	8.5
	SE	2.3	2.0	2.1	2.0
Rough and tumble	*M*	5.4	2.9	4.7	4.1
	SE	1.7	1.5	1.6	1.5
Fantasy	*M*	10.8	8.0	5.8	1.5
	SE	2.4	2.1	2.2	2.1
Games with rules	*M*	5.1	0.0	17.3	10.5
	SE	3.4	3.0	3.2	3.0

[a]*Konomia*, 4–6 years old; *Otyaro*, 7–12 years old

Groups of up to 30 A'uwe-Xavante children play "jumping into the river." The youngest occasionally try to jump in, but most of the time they are content to watch, apparently fascinated by the boldness of the oldest, who invent many alternative ways to jump into the water (Nunes, 1999). Puddles also promote play; during the rainy season the Kaingáng (Pereira, 1998) and A'uwe-Xavante children can be seen happily slipping in the mud (Nunes, 1999).

Climate variables have been considered as determinants of the level of exercise play in children. Cullumbine (1950) found low levels of exercise play in peoples living in tropical climates and Pellegrini, Horvat, and Huberty (1998), comparing different temperatures, also report lower levels of exercise play during warmer periods. Parakanã children live in a tropical environment; however, their easy access to a river may explain their relatively high levels of exercise play.

In urban samples, exercise play has been found to follow an inverted-U developmental curve, peaking around 4–5 years (Eaton & Yu, 1989; Routh, Schroeder, & O'Tuama, 1974). In their review, Pellegrini and Smith (1998b) report that exercise play accounts for 7–10% of observation time in 2- to 4-year-olds both in day care settings and at home; and at 5–6 years of age, it accounts for about 13% of both in-home behavior and outdoor behavior during school recess. McGrew (1972) and Smith and Connolly (1980) found that exercise play (defined as chasing, jumping, climbing, etc.) accounted for 20% of all observed behavior in school.

Gosso found that exercise play, as defined by McGrew (1972) and Smith and Connolly (1980), among Parakanã children at Paranowaona village accounted for 24%, plus or minus 8%, of observation time. This value is higher than those reported in most of the studies mentioned above. Table 9.1 shows results by gender and age (the Parakanã Indians call 4- to 6-year-old children *Konomia* and 7- to 12-year-old children *Otyaro*). On average, boys engaged in more exercise play than girls (as expected from the literature, which reports reliable sex differences; Pellegrini & Smith, 1998b), and, on average, *Otyaro* engaged in less exercise play than *Konomia*—a result compatible with an inverted-U developmental curve. However, the effects of sex and age did not reach statistical significance.[2]

EXERCISE PLAY WITH OBJECTS

Objects are often present in physically vigorous play activities that may demand ability and balance. For example, forager children walk high above the ground, standing on stilts (named *My'yta* by the Camaiurá), which are very common among Amazonian Indians. The Camaiurá build them with wood and straps of *embira* (a tree from the Amazonian forest) that serve to hold their feet firmly in place. They use a knot to adjust the strap to the child. This toy can also be used to simulate footprints of birds and other animals (Moisés, 2003).

Draper (1976) and Wannenburgh (1979) described !Kung children playing *zeni*. This is a solitary game practiced by both boys and girls, from about 6 to 15 years old. The *zeni* is a feather attached to a weight (a pebble or an uncracked *mongongo* nut) with a leather thong (about 15 centimeters long). The player hurls the *zeni* using a stick. The weight falls first, and the feather flutters behind, acting as a kind of parachute. The objective of the game is to reach the *zeni* and strike it again before it hits the ground. As it spins slowly to the ground, children dart beneath it to flick it up again. Draper emphasizes the fact that each player seems to be perfecting his or her technique, and does not seem to be trying to beat the rest. She never saw any child counting how many times each player could hit the *zeni* without letting it fall. The participants would play again and again, motivated by the pleasure of playing and to improve their own performances, without any competitive interest. The /Gwi believe that, were it not for the *zeni*, humankind would still be living in darkness. According to a /Gwi legend, Pisiboro, an ancestral hero, made a *zeni* picking a feather from the wing of a *korhaan*, a grassland African bird, and weighting it with a burning coal. He flung it twice into the sky, and it floated back to earth. Than he used a stick to hurl it a third time, and it flew so high that it remained in the sky, becoming the sun.

Children of the !Ko, Himba (Namibia), and Eipo peoples were observed playing ball (Eibl-Eibesfeldt, 1989). Cooper (1949) reported the widespread occurrence of different types of play with rattles and balls among Indians from many Central and South American societies (Cooper, 1949).[3] Balls are frequently made out of maize leaves or rubber extracted from *Hevea brasiliensis.* Games can vary, from simply throwing the ball from player to player, as found among the Indians from Tierra del Fuego, to kicking the balls, as the women from the Pampas do, up to more complex games with targets.

OBJECT-ORIENTED PLAY WITH FINE MOVEMENTS

Objects may also be the focus of playful activities that do not involve intensive exercising. The Camaiurá and the Yawalapiti of Mato Grosso (Brazil), for example, make spinning tops out of their regional fruits—toys that spin around their points when the children twist them (Moisés, 2003). The Deni, a small group of the Aruak family, which inhabits the basin of the Purus river, makes tops out of *tetiaru* (*Anthodiscus amazonicus*), an Amazonian fruit, by extracting the heart of the ripe fruit and inserting a wooden spindle through its center; they also drill two lateral holes, so that the tops whistle as they spin (Ribeiro, 1987). Camaiurá children build a buzzer out of a disk of calabash, which rotates when a cord of Tucum (*Astrocaryum aculeatum*, a palm from the Amazon region) is wrapped and unwrapped around it.

Forager children are also frequently seen playing with bows and arrows and slingshots. This behavior was observed among boys of the Guayaki (Clastres, 1988) and the Parakanã from Pará (Gosso & Otta, 2003). It is popular knowledge in the Amazon area that such bow-and-arrow play is found in many other Brazilian tribes, located far away from one another, such as Sataré Mawé and Guajá, Carajá, Urubu-Kaapor, and Guajajara. In general, this play is exclusively masculine. Four- or five-year-old Guayaki boys receive a small bow, adapted to their size, from their fathers. From that moment on, they begin to practice shooting arrows. As they grow, they are given bigger and more efficient bows and arrows (Clastres, 1988).

!Kung boys observe, with great interest, adults making spears and arrow points. Draper (1976) observed a man hammering and shaping a metal for several arrow points, and as he worked, two 4-year-old boys (his son and grandson) sat on his legs and attempted to pull the arrowheads from under the hammer. The man did not get angry or chase the boys off, and they did not give heed to his warnings to quit interfering. When the children's fingers came close to the point of impact, the man waited until the small hands were a little farther away to continue hammering.

Figure 9.5 shows three Parakanã boys practicing the use of bow and arrows. Gosso and Otta (2003) observed three variations on the bow-and-arrow play of the Parakanã Indians. The simplest form involves throwing arrows upward, without a target. This modality was observed only among 8-year-old boys:

> Arara'ywa, an 8-year-old boy, throws his arrow and goes searching for it. He finds it and throws it again. He runs to the place he had aimed at. He looks for the arrow and laughs when he finds it, runs to get the arrow and throws it once again, returns smiling to get his arrow. He throws it again and goes to get it, followed by Ma'apyga (10 years old, M).

Another modality involves practice on a static target:

> Kyryry'ia, an 8-year-old boy, casts his arrow and smiles. He gets his arrow, looks at a potato that lies at a distance of a foot from the arrow. He throws and hits it. He takes the arrow and aims again but at a greater distance and misses twice. He tries again and succeeds. He walks towards his home carrying his arrow still with the potato impaled on it.

FIGURE 9.5. Parakanã boys playing bow and arrow. Photograph by Y. Gosso.

Following and hunting a live animal was observed in boys from 10 years old and more:

> Waripa, a 10-year-old boy, goes into the forest with his arrows and catches sight of a lizard. He walks into the forest, fires an arrow, but fails. He follows Ape'ea (15 years old, M) searching for another lizard. He follows a trail in the forest. . . . He sees a lizard, but lets Ape'ea fire at it. He leaves the forest running and shaking off ants from his back. He continues following Ape'ea. Circles a hut together with Ape'ea, searching for another lizard. Enters the forest again still following Ape'ea.

Only once was a 12-year-old girl at Paranowaona village seen playing with a small bow and arrows. She was near the forest accompanied by a 4-year-old boy (Gosso & Otta, 2003).

The fact that fathers give little bows and arrows to sons and mothers give little baskets to daughters should not be taken as deliberate sex-role teaching or fostering in any reasonable sense. They are merely following a traditional usage in a very casual way. Adults do not give any indication of being worried about the psychological future of their children. Whether or not their children will become effective adults is not an issue.

CONSTRUCTION PLAY

One form of object-oriented play is construction. During the rainy season, A'uwe-Xavante boys and girls of the Nambukurá village spend hours building small houses with mud. Older children use sticks up to 20 centimeters long to create the house's structure. After the walls have been built, internal divisions are added in order to separate compartments: places to sleep, to cook, etc. Up to three houses may be built next to each other. Nunes (1999) observed that a group spent around six hours in house building with mud and sticks. Younger children play with such structures, once they are abandoned, because they find them difficult to build.

!Kung children build little huts (Shostak, 1976, 1981), and !Ko Bushmen, Himba (Namibia), and Eipo children play sand games. These children use some objects in their activities, and there are gender differences in the frequency with which sticks and stones are used in play, favoring the boys (Eibl-Eibesfeldt, 1989).

Among Parakanã Indians, Gosso found a gender × age interaction effect on construction play.[4] The *Konomia* (age class 4–6 years) girls were more engaged in constructive play than boys, whereas the opposite occurred within the *Otyaro* (age class 7–12 years; Table 9.1). Older girls, in general, do not engage in construction play, whereas older boys become even more interested in this kind of activity. The differences between boys and girls should be understood in the light of some details of the different kinds of play in the broad category of construction play and in the wider context of gender identity. Some play was seen only in girls: building with sand, straw weaving, and body painting. Except for sand building, these are feminine activities in Parakanã culture. The boys engaged in digging, house building, and other forms of play with wood and stones—all typical male activities. Even when playing with clay, done by both boys and girls, there were gender differences regarding the objects that were made. The human figure, for example, was made only by girls.

For some reason—earlier gender identification or the nature of the tasks or more contact with women than that of boys with men—little girls are ahead of boys at construction play, and they may lose interest earlier as a result of competition with other tasks, such as taking care of very young children and helping adult women with their work. Gender identification among the Parakanã seems to be the result of a wide and powerful network of relationships, subject to little or no risk of failure. It is something like a two-way road: The interests and preferences of the children are duly responded to by adults, and vice versa, on easily shared psychological ground. It is not a matter of concern (or even awareness), and it is certainly not the result of anything that could be described as deliberate guidance.

Bjorklund and Pellegrini (2002) report several studies with urban samples that show quantitative gender differences in construction play. Boys tend to build more structures than girls (Caldera, O'Brien, Truglio, Alvarez, & Huston, 1999) and are more likely to manipulate objects, taking them apart and attempting to put them together again (Hutt, 1972; Sutton-Smith, Rosenberg, & Morgan, 1963).

Among foragers, there are qualitative gender differences in the construction play interests, in addition to the quantitative gender differences reported. Gikuyu boys from Kenya make axes, spears, slings, and bows and arrows, like their fathers, whereas girls make pottery (for cooking real or imaginary food), clay dolls, and baskets of plaited grass (Leacock, 1976).

The use of mud to shape dolls and animals is also common among South American Indians, who find raw material at river banks. The Kaingáng (Pereira, 1998), Mehináku (Gregor, 1982), and Parakanã children (Gosso & Otta, 2003) mold turtles, alligators, jaguars, dolls, and bowls with mud. Figure 9.6 shows a group of Parakanã children shaping objects with clay. All human figures made by Parakanã children have clearly defined genitals. Gosso observed only girls modeling human figures, both adults and babies, and they seemed at ease to shape and to exhibit the dolls' genitals.

Mehináku children use interweaved threads to make a "cat's bed" (figures or simply a beautiful puzzle of threads); in addition to their hands, they use their feet and teeth in this activity (Gregor, 1982). The Camaiurá make figures, such as bats and fish, with threads they interweave with both hands; at least 10 different figures were observed being created by the Camaiurá (Moisés, 2003).

Making baskets out of palm leaves is another common form of play among the Parakanã (Gosso & Otta, 2003), Canela from Maranhão State, Xerente from Goiás State, Txicão from Mato Grosso State, and Waimiri-Atroari and Tucano from Amazonas State, as can be seen in the Indian handicraft collection of the *Museu do Índio do Distrito Federal* (Museum of Brazilian Indians in Brasilia). In the Parakanã group, this activity was observed exclusively among girls. These girls take a big knife and go into the forest in small groups; they cut green palm leaves and carry them to a specific site, generally an opening in the forest or an abandoned house. There they completely weave the baskets (*peyras*) or just interlace some leaves (Figure 9.7). These *peyras*, as well as other weaved materials, were only used in playful contexts. Much chatting takes place during the activity. This play was mainly observed among small children, maybe because girls older than 8 years weaved their baskets and other tools for real use.

FIGURE 9.6. Children modeling clay. Photograph by Y. Gosso.

FIGURE 9.7. Girl pretending to make a basket. Photograph by Y. Gosso.

Another playful activity, exclusively feminine, is body painting. This activity should be considered playful because it occurs outside of adult ritualistic contexts. Some Parakanã girls were seen painting their peers or their own bodies with ink extracted from urucum bushes (*Bixa orellana*). The fruits of these bushes are small and reddish in color, with a soft peel that allows easy access to their seeds. The plant is found in neighboring locations around the village. When the seeds are rubbed or smashed by means of a small twig, they release a red ink commonly used by the Indians for body painting. In the Northern part of Brazil, the juice of this fruit is commonly sold in markets and used by housewives as a natural colorant for cooking (Gosso & Otta, 2003):

> Iara, a 4-year-old girl, watches Mameia (5 years old, F) with a bowl filled with urucum fruit on her lap. She opens a fruit, takes the seeds in her hands, and pastes them on her face. She gets up, and sits in front of Moropyga (4 years old, F) who has just arrived bringing more urucum. She separates more seeds and rubs them in her hands. She gets up and runs toward an urucum bush. She gets under the bush searching for more urucum, looks at the observer and smiles. Wewe (5 years old, F) takes a bunch of urucum for her and another for Iara. Both girls return running to the initial point, when Y'yma (a visitor from another village, 5 years old, F) asks to paint Iara's body. Iara sits with her back toward Y'yma and stays still while Y'yma paints her back, and later on, her face.

SOCIAL CONTINGENCY PLAY

Forager children engage in social contingency play when they are still very young, taking turns with their partners. Among !Ko Bushmen, Himba (Namibia), and Eipo people, less than 1-year-old children were observed tirelessly playing out dialogues of give and take (Eibl-Eibesfeldt, 1989). There are no extrinsic rewards, and they appear to get high excitement and satisfaction from the activity itself.

Various forms of social contingency play are observed as children get older. Gregor (1982) reports two games among the Mehináku, which combine social contingency with motor exercise. One of them requires specific abilities employing a rope. The children get pieces of rope and exhibit several skills with them. They encircle the neck of a child with a rope, making a false knot, and suddenly pull the ends of the rope. If the knot was correctly tied, it releases. Children like to demonstrate these skills in public. Another form of rope play that requires motor coordination and attention is called the "wasp game." Older children carefully make a spiral on the ground with the rope, representing a wasp's nest. A younger child tries to repeat the design, tracing with his or her finger, while others watch with their hands full of sand. If the child does not draw an exact copy of the original line, the others scream and toss sand on him or her.

The Camaiurá children play "Where is the fire?" They dig two holes in the sand and connect them with a tunnel. One child places his or her head in one of the holes and is completely covered with sand by the other children. The child breathes through the tunnel as the others provoke him or her with mockery and ask "Where is the fire?" The child underground has to indicate the correct position of the sun at that moment. The game finishes when he or she gives the correct answer (Moisés, 2003).

"Hide-and-seek" forms of play were seen among the Mehináku (Gregor, 1982), the Parakanã (Gosso & Otta, 2003), and various other Indian peoples from South America (Cooper, 1949).[5] A variation of the hide-and-seek play is found among the Camaiurá and the Sucuri. On a stump or big rock over a lake, one of the children, "the fisherman," waits for a partner to dive into the water and pass nearby; the "fisherman" then jumps into the water and tries to catch him or her. The children stir the mud at the bottom of the water to make it murky and create difficulty for the "fisherman" to locate his or her prey (Moisés, 2003).

In the hide-and-seek play of the Parakanã Indians, children of different ages participate. Generally older children help the youngest to hide and have fun watching them. Those who hide do not always remain in the same place. They laugh and change their hiding place. The child in charge of searching waits for a sign from the partners to begin the search (Gosso & Otta, 2003). According to Peller (1971), the hide-and-seek play follows a different formula depending on the age of the children. Whereas for older children it is important to be smarter than their partners, for younger children this play involves assuring the return of those who leave; by producing the separation and the reencounter, the hiding child can dominate the emotions involved in such situations. Younger children hide, then frequently reappear before being found.

As can be seen in Table 9.1, social contingency play among Parakanã Indians is more frequent in girls than in boys.[6]

ROUGH-AND-TUMBLE PLAY

A special form of social contingency play is rough-and-tumble (R&T) play. !Ko-Bushman from Central Kalahari display a kind of playful aggression; patterns that would lead to subordination and a severance of contact in a serious context are performed during play fighting. Although the motor patterns involved are often difficult to distinguish from real aggressive acts, laughter and smiling function as signals indicating that the interaction is not aggressive but playful. The roles of attacker and defender, or pursuer and pursued, change freely (Eibl-Eibesfeldt, 1974). !Ko Bushmen, Himba (Namibia), and Eipo boys also like to play at fighting and hunting (Eibl-Eibesfeldt, 1989).

!Kung children play games of chasing and striking dogs, insects, and other available living creatures, including people (hitting people is laughed off and not discouraged)—all these being essential behaviors during hunting (DeVore & Konner, 1970). !Kung boys play with ostrich chicks; when the play becomes more boisterous, the chicks run away. What begins as play may end in a kill; boys have been observed running down chicks and killing them with sticks and stones (Wannenburgh, 1979).

Playful chase-and-fight games are also common among children of South American Indians. Camaiurá children play wrestling in the water; a child stands on the shoulders of another and fights against a similar pair. The children on top hold hands and try to pull their opponents down (Moisés, 2003). A'uwe-Xavante children chase each other (Nunes, 1999, 2002); Kaingáng girls play "chase" while boys play "wrestling" (Pereira, 1998); and the Mehináku play *yanomaka* (jaguar), in which a child ambushes unsuspecting partners. Children from the Camaiurá village like to play *Kap*, which means *wasp*. Both boys and girls divide into two groups: One group represents the people of the village and the other builds a wasp nest in the sand (a hill in the middle of a circle). While the children build, they imitate the buzzing sound of wasps. Then the first group discovers the nest and tries to destroy it, while the attacked "wasps" react and pursue the villagers. Wasp stings are represented by pinches, and the "wasps" only cease when the villagers stop running (Moisés, 2003).

Xavante boys from Mato Grosso (Bastos, 2001) form two teams, painted red and black, each with the symbol of his clan on his face. The two boys who will fight one another are chosen by the oldest boys of each clan. The children hit each other's arms with the roots of a strong grass until one of them gives up.

A turbulent form of play found among the Camauirá is the *Jawari*. One of the participants throws his spear toward a fence made of poles, trying to knock down the poles. The other players are lined up behind the fence and are not allowed to leave. As the "pitcher" knocks the poles down, the other players become easy targets and must then divert the spears, without moving their feet. Some are hurt during this game.

The *mocareara angap* is a type of pressure weapon, made by the Camaiurá Indians, with a bamboo tube that uses a native fruit called *pequi* (*Caryocar brasiliense*) as pellet. With this weapon the Indians play hunting: One boy hides in the forest and the other searches, shooting as soon as he finds the hidden one (Moisés, 2003).

Gosso observed R&T with peers among Parakanã children (Gosso & Otta, 2003). Because observations were made of a nonurban sample and outside the conventional school settings in which most R&T studies were conducted (Boulton, 1992; Humphreys & Smith, 1987; Pellegrini, 1984), high rates of this form of play were expected. This was not found. Although the children were on their own, and there was no "school policy" discouraging R&T, it only accounted, on average, for 4% ± 4% of observation time. This value is not very different from those reported for urban samples: 3–5% of observed classroom behavior during the preschool period, and 7–10% at ages 6–10 years (for a review, see Pellegrini & Smith, 1998).

Eibl-Eibesfeldt (1989) reported sex differences in playful pursuit and fighting/competitive games among !Ko, but did not discriminate the age of the participants. Playful pursuit accounted for 18% of male and 4% of female game activity. Fighting and competitive games accounted for 15% of male and 0% of female activity.

Although, on average, Parakanã boys engaged in more R&T than girls (Table 9.1), as expected from the literature that reports reliable sex differences (DiPietro, 1981; Pellegrini, 1989; Smith & Connolly, 1980), the main effect of sex did not reach significance.[7]

If R&T were taken as a less comprehensive concept, restricting it to include only actual play fighting with intense bodily contact, this kind of play would be nearly absent in Gosso and Otta's (2003) records. Parakanã children, both boys and girls, do know what fighting means, but both play fighting and actual fighting are very rare, indeed. In a few instances, Gosso saw play fighting merge into actual fighting, with cries and tears in the end. Surprisingly, adults were not in the least alarmed, nor did they do anything about it. Teenagers and women laughed about it without interrupting their activities. It is our view that play fighting, so important in so many mammals, may have lost nearly all of its relevance for humans. Its abundance in other mammals relates to the adult male's need to fight for status, females, territory, and food (e.g., Byers, 1984, for ungulates). In animals there seems to be a direct relation between adult agonistic competition among males and the performance of play-fighting movements early in life. Polygynic species exhibit more adult fighting, sexual dimorphism, and juvenile play fighting. Byers (1984) suggests that selection pressures for early rehearsal of adult competitive patterns probably did not appear until social groups and the potential for polygyny appeared. In our view, at some early point of human evolution, those benefits resulting from superior fighting abilities must have declined. Vertical hierarchy gave way to horizontal cooperation among males (and females). Access to females ceased to be the direct or indirect result of fights. A unique food-sharing pattern, devoid of priorities and disputes, made obsolete the advantages of superior fighting abilities.

FANTASY PLAY

Children of all forager groups studied exhibit fantasy play. !Kung children play at being hunters striking animals with their make-believe arrows, then taking leaves and hanging them over a stick, pretending that these are meat (Shostak, 1976, 1981). Wannenburgh (1979) observed a !Kung boy making himself a horse and rider from balls of dry dung inspired by the sight of men hunting giraffe on horseback. !Ko children play out the trance dance (Eibl-Eibesfeldt, 1989). !Kung children also play trance dance; they sing and dance, and the boys make-believe that they are curing the girls. Children also engage in sexual play with one another (Shostak, 1976, 1981).

Children's use of dolls that are treated as babies in symbolic games is very common among many foraging communities. A melon is a substitute for a doll among the !Kung. A banana blossom is carried as a doll in a sling by Yanomami girls. A banana is cuddled as if it were a baby by a 2-year-old Yanomami girl, still a baby herself (Eibl-Eibesfeldt, 1989). Cooper (1949) noted that dolls are present in the play of children in many different societies.[8] As to materials of which dolls are made, he listed wood, clay, wax, bones, and straw. An exhibit sponsored by FUNAI (Brazilian National Foundation for Indians) displayed dolls made of cloth (Guajajara), ceramics (Carajá), wood (Carajá and Canela), chestnut cortex (Caiapós and Tucuna), and turtle shell (Caiapós). Our Parakanã Indians make dolls of clay.

During the rainy season, A'uwe-Xavante (Nunes, 1999, 2002) and Parakanã children (Emídio-Silva, personal communication) use sticks to make trails on the soil. Objects are also used as vehicles in imaginary trips (Figure 9.8). Older children are seen pushing younger ones in boxes, as if these were vehicles passing through the village.

Nine or ten-year-old Guayaki girls receive miniature baskets from their mothers. Such baskets are not actually used for transporting things, but the girls imitate their mothers' activity with them (Clastres, 1988). Boys and girls tend to selectively imitate their own sex. Xikrin girls play house, caretaking, gathering, and potato cooking. Boys play hunting lizards with bows and arrows, and house building (Cohn, 2002). Yanomami girls offer their breast in play to smaller children (Eibl-Eibesfeldt, 1989). Among the Xocó Indians from Sergipe, girls play baby caretaking; they pretend preparing food and other domestic activities, whereas boys pretend they are horse riding, canoeing, or ox-cart driving (Bichara, 1999). The Kaingáng girls play with dolls made out of corncob, and boys use pieces of wood as vehicles (Pereira, 1998).

FIGURE 9.8. Parakanã child pushing a plank in a puddle. Photograph by Y. Gosso.

Gregor (1982) narrated four fantasy plays of the Mehináku children in great detail: "children of the women (*teneju Itãi*)," "marriage," "chief," and "shaman (*yetamá*)." These plays reproduce complex scenarios full of ritualistic elements.

"Children of the women" (*teneju Itãi*) is a kind of fantasy play that involves both boys and girls from 5 up to 12 years old; it is performed at a good distance from the village, in places where they cannot be seen by their parents or siblings. The make-believe begins by forming couples, as if married. Each couple makes a "child" from a block of earth, on which they shape arms, legs, facial traces, and genitalia. They rock the "babies" in their arms and talk to them. The "mother" carries her "child" on her hip and dances with her (imitating what her mother does with younger siblings). After the "parents" have played with their "child" for a while, the "baby" falls asleep or dies. If the "baby" dies, the "parents" cry, dig a grave, and bury their "baby." Then all "mothers" kneel in a circle, with their heads lowered and their arms on their partner's shoulders, lamenting and crying for the lost "child." Gregor (1982) stipulated that when he saw such fantasy plays, the children seemed to be having great fun.

"Marriage" is a kind of fantasy play in which the children enact the complete matrimonial ceremony, and after the nuptial ritual, they lie in hammocks as if they were real couples. The "husband" leaves to hunt or fish and returns with leaves that represent fish or monkeys. The "wife" cooks food and distributes it among the men. This game presents some variations, such as the one called "jealousy," when boys and girls find "lovers" while their spouses are not home. When the spouse discovers the infidelity, he or she angrily beats the unfaithful "husband" or "wife," while the lover runs off.

"Chief" is a kind of fantasy play with the tribe's politics as its theme. The children divide into two tribes that represent the Mehináku and one of their neighbors. Each tribe builds its own village, and in each, the "chief" gives a speech to his people at the center of the square. The "neighbor tribe" visits the "local tribe" for a formal interchange of goods, and, as in real life, the "chiefs" of each village exchange presents and the men fight. The children interchange their clothes and other personal possessions.

During the "shaman" make-believe play (*yetamá*), one of the children fakes being sick and the "parents" leave to request the help of the best "shaman" in the village. Minor "shamans" sit around the patient and pretend to smoke cigars. Once in a while, they suck the patient's body to remove the disease. When the strongest "shaman" arrives, he smokes, fakes fainting, then recovers and runs around the village searching for evidence of witchcraft. After a small time interval, he returns with a piece of string or wood, saying that these were made by a wizard to produce the sickness.

There is also a kind of fantasy play in which the children pretend that they are secluded. This play represents a ritual passage from adolescence to adulthood, during which they spend a period of time separated from the rest of the people. There is a variant in which a girl breaks the privacy of the men's house, and the boys, to get even, pretend to abuse her sexually. Gregor (1982) commented that these fantasy plays of the Mehináku children are not just a replica of the adult society, but a surprisingly exact copy of it. The accuracy of the roles represented reveals not only that the children know about public matters of the village but also the private world of their parents.

Among the Parakanã Indians, Gosso also observed fantasy play that reproduces adult habits. One day, a group of boys sat in the house where the men typically held their night encounters (*tekatawa*) to tell stories, plan hunting strategies, decide marriages, and discuss other matters of interest for the community. Just as the adults, only boys participated; they simulated a meeting in which they used a piece of wood to pretend they were smoking a Parakanã cigar (*petyma*; Figure 9.9). They not only "smoked," but also sang and danced.

The richness in details with which indigenous children play the roles of the adults of their society is remarkable. The characteristic social straightforwardness of indigenous groups allows their children to gain a more complete social vision than that of children in industrialized societies (Gregor, 1982). The observations made by one of us (M.L.S.M.) of make-believe among preschool children of São Paulo city, especially among boys, demonstrates this point. The father in the urban boy's fantasy play appears in a car just coming from, and going to, work. The children richly reproduced the activities of their TV heroes but poorly imitated the type of work done by their parents (Morais & Carvalho, 1994).

Gosso found that Parakanã children of the *Konomia* age class (4–6 years old) exhibited more fantasy play than those of the *Otyaro* age class (7–12 years old)[9] (Table 9.1), a result compatible with an inverted-U developmental curve also found among children in industrialized societies (Fein, 1981).

Gosso, Morais, and Otta (2002) compared 4- to 6-year-old children of five different cultural groups: Parakanã Indians, seashore, low socioeconomic status (SES) urban, high SES urban, and mixed SES urban. The seashore children are not Indians; they are culturally similar to the Brazilian population who live in small villages. The authors found that a lower proportion of Parakanã Indians' and seashore children's playtime was dedicated to make-believe play. High SES urban children engaged in a greater propor-

FIGURE 9.9. Boys pretending to hold *tekatawa* (men meeting). Photograph by Y. Gosso.

tion of pretending than both Indians and seashore children, whereas low SES and mixed SES urban groups were in between the two. It is known that upper- and middle-class urban children play more make-believe than low-class children (Fein, 1981). In addition, little symbolic play, characteristic of Indians and seashore children in our study, was also found by Martini (1994) in the Marquesas Islands in Polynesia and by Gaskins (2000) among Mayan children. The low frequency of pretending in Indian, seashore, and low SES urban societies may be explained by the predominance of simple communication codes in these societies. Insofar as they are more concerned with material survival issues, their way of rearing emphasizes concrete and immediate solutions to their daily life problems, instead of strengthening symbolic and abstract thought.

The emphasis of Indians and seashore cultural groups on practical solutions to daily life problems is associated with precious developmental opportunities: deep contact with nature and native forests, hills and high trees to climb, rivers, sea and wild animals to master. Furthermore, the children live under less adult supervision, and almost all adult world objects are available to them, without prohibitions. The highly challenging environment in which they live might reduce their tendency to look for fulfillment of desire through fantasy. This interpretation is consistent with the finding that more fantastic themes were practically absent in the play of Indian and seashore children (Gosso & Otta, 2003; Morais & Otta, 2003).

GAMES WITH RULES

Games characterized by more structured rules than play activities have been described in foragers. Roles are ritualized, and predictable and predetermined scenes are represented in repetitive cycles of actions (Parker, 1984).

The "hawk game" is found among the Canela Indians (Bastos, 2001). One child plays a hawk and the others stand on a line that begins with the tallest child. Each child in the line strongly holds the one ahead. The "hawk" is released and screams "*Piu*" ("I am hungry"). The first child in line opens his or her legs, bends down so that his or her head is between the legs, and looks back, pointing his or her arm to the child behind and asking, "*Tu senan sini?*" ("Do you want this?"). All the children in line repeat this question. The "hawk" answers "*É pelá*" ("No") to all except the last one, to whom the "hawk" says "*Iná*" ("Yes") and then begins to chase him or her. The group, without letting go of their physical contact, try to trap the "hawk." If the "hawk" catches the child, he or she will take him or her to his or her nest. The game continues until the "hawk" gets all the children, following the order of the line.

The "jaguar game" is similar to the hawk game. It is found among the Macuxi Indians from Rio Branco (Brazil) and other South American Indians (Cooper, 1949). A group forms a line in which every child holds tight the waist of the child ahead. Whoever is at the beginning is the "jaguar": He or she jumps from one side to the other, makes noises, and moves his or her "tail." Suddenly the "jaguar" jumps and tries to grab the last child. The line moves to avoid the capture. When the "jaguar" gets the child, he or she takes him or her to his or her cave and returns to repeat the game, until he or she has caught all the children.

In the forests around the village, the Mehináku children divide into two groups to play the "monkey game" (*paí*). The "monkeys" climb the trees and jump on the branches, whistling and making noises, while the "hunters" throw sticks at them, trying to make them fall to the ground (Gregor, 1982).

"The pumpkin thief" is a common game among the Canela children (Bastos, 2001). The children who play the "pumpkins" in the field squat in one place. The "owners" of the field stand a bit aside, watching over the "pumpkins." Two children playing "elders" arrive to buy pumpkins: They walk hump-backed and with the help of a cane. When the "buyers" get near the field, they ask the "owners" if they want to sell "pumpkins." The "owners" say no, and the "elders" pretend to leave. The "owners" go away, and the "elders" return to steal the "pumpkins": They softly hit the "pumpkins" on their own heads to see if they are ripe. Then they hide them.

When the "owners" return and find the pumpkins gone, they protest to the "elders," who say they do not know where the "pumpkins" are. A fight begins. One of the "owners" fights with the "elders," while the other looks for the "pumpkins" and finds them. The "elders" get angry and decide to steal the "pumpkins" again, and the game continues, like this, until the children get tired.

Cooper (1949) observed some games with well-elaborated rules, such as hockey and *pillma* (a game in which players form a circle; a ball is thrown beneath the thighs, and the players try to keep it in the air). *Pillma* is present among Araucanian, Pampean and Patagonian people.

Widely disseminated among the Indian groups east of the Andes mountain chains and north of the Pampas are log races, hockey (similar to the one we know, with canes and rackets), hoop and pole (in which a spear is used to hit a moving target), corncob darts, bean shooters, and tops.

Ball games are played between teams of youngsters and adults and between tribes and villages. In most tribes, the game's objective is to hit targets placed at the end of the field. Among the Witoto, Macuxi, Patamona, Xerente, Mojo, and Parakanã, the ball is thrown with both feet and hands. The Apinayé hit the ball with something like a bat. Games with marbles are also common among groups that live in the Amazon region, Orinoco, Guyana, East of Bolivia, and Brazil. Players throw the marbles (e.g., seeds of maize) and hit them back with the palm of their hands.

Among the games described by Cooper (1949), the only ones with material prizes for the winners were ball games, *pillma,* and hockey. The prizes were baskets full of corn, necklaces of beads, arrows, or other things that the players had at home. It is probable that these games have existed since pre-Columbian times, and that they had links with Mexican and Antillean games.

The Camaiurá play *ui'ui*. In this game, a fine and resistant thread made of fiber from a special type of palm tree (*buriti*) is partly buried horizontally. The organizer of the game covers the thread without letting the others see its direction on the sand. The winner is the person who discovers this direction. Even though the organizer keeps one end in his or her hand, this does not mean that the thread is extended straight; usually, he or she makes a curve to deceive the participants. The organizer makes rapid forward movements, letting the tip appear briefly, then disappear again. What complicates the game is the fact that there are several false *buriti* tips scattered in the sand (Moisés, 2003).

Adugo is the name of a board game found in the Bororo and Pareci tribes from Mato Grosso (Brazil). The board is drawn in the sand; one stone represents a feline (a jaguar) and 15 other stones represent dogs placed at different points. One player moves the "jaguar" to eat the "dogs"; it jumps over squares, as occurs in checkers. The opponent moves the "dogs," trying to surround the "jaguar" and stop it from gaining ground (Moisés, 2003).

Dice are another game found among South American Indians. The dice are made in different shapes of wood or bone: a quadrangular prism among the Canela; a six-faced hexangular pyramid among aborigines from Peru and Ecuador; a five-faced prism or an oblong pyramid among some Ecuadorian groups. The dice are thrown from a small height and a score is kept by counting marks on the soil (Cooper, 1949).

Variations on dice games have been found, such as the *lligues* in the Araucanian area and the *tsúka* in Chaco, that are aboriginal games. *Lligues* is played with 8–12 grains that are painted black on one side; the other side is painted white or left natural. The player shuffles the grains and throws them to the ground, counting the points in relation to the colors exposed. Splinters are used to register the points. In a game called *Tsúka,* four wooden dice are used; these are almost 10 centimeters wide, convex on one side and concave or flat on the other. The player takes two dice in each hand and throws one pair against the other so that they touch when they reach the ground. Points are gained when the convex side appears (Cooper, 1949).

Gosso found that Parakanã children of the *Otyaro* age class (7–12 years old) spent higher proportions of the observation time playing games with rules in comparison with the *Konomia* age class (4–6 years old)[10] (Table 9.1). Rule games were more frequently played by older children, whereas younger children preferred symbolic play (Gosso & Otta, 2003).

The increase in the frequency of games with rules and the reduction of the frequency of fantasy play as age increases has also been found for samples taken of children in urban areas of different countries: Brazil, Korea, United States, France, Switzerland, and Sweden (Piaget, 1962; Sinker et al., 1993).

In her study of Parakanã Indians, Gosso also found a marginally significant main effect of sex. Boys were observed playing games with rules more frequently than girls (Table 9.1). It is interesting to note that in his classic study on the game of marbles, Piaget (1965) reported that boys engaged in games with rules more frequently than girls and also that their rules were more complex (Parker, 1984).

CONCLUSIONS

Hunter-gatherer children play. They neither hunt nor gather, they do not build houses and shelters, they do not cook or clean. In fact, the life of children, at least under 7 years of age, is mostly, if not solely, a playful life. From early morning to bedtime, they play, typically all together. So far, this pattern is not surprising, given that we are mammals, and most, if not all, comparable mammals display the same pattern. Wherever the reproductive strategy allows for several young to develop together, play is their full-time activity, except for being nursed; this strategy also applies to humans, of course. One important difference between humans and other mammals is that human children, in the hunter-gatherer pattern, have plenty of time to play together in the absence of adults, and in a larger group than that of siblings. Another difference is that human children have poor preemptive and reactive behaviors against predators, for they are by far less quick and functional in the presence of real threat.

Whatever the functions of human play, it is reasonable to believe that some of them are shared with other mammals. It will be in the differences that we may find some insight into the phenomenon of human play. For instance, if it is true (as observations of foragers strongly suggest) that the human pattern involves allocating a very large part of the day to playing, we could deduce that human adults are thereby free to do what they have to do, such as taking care of the highly demanding newborn, hunting, gathering, building, tool making, and so on. In other primates, adult females seem to be much less free of their young. For example, Clark (1977) reports that young chimpanzees remain with their mothers for 9–12 years, and their mothers are their only companions for most of the time. So, at least in part, one function of human play could be to enhance the effectiveness of adult behavior. By being able to gather together in a group that is capable of remaining on its own all day long, children relieve their parents from the burden of caring for them. One feature of the group of children is the presence of very young members; it is remarkable to see 7- and even 5-year-old girls taking care of 2-year-olds.

However, going beyond those three basic differences, and trying to understand the content of human play, perhaps the most striking peculiarity is that children seem to live, as they play, in a special culture that is different from the adult culture. Seemingly, we should develop the concept of a culture within the culture, namely, children's play culture. Such children's play culture, of course, is related to the adult culture, and neither children nor adults remain totally absent from either culture. Children play inside their own culture but are also in contact with the adult culture. And adults partly understand and occasionally enter their children's play culture. So children play at culture; *culture* is the name of the game, because they are innately ready to exhibit cultural traits (Bussab & Ribeiro, 1998), but the adult culture is beyond their full reach and physical abilities. As a consequence, a peculiar cultural environment emerges in which they spend their first years of life.

The hunter-gatherer cultures set a very clear turning point from childhood into adulthood, thus separating the two cultural settings. Those two settings are not of the same nature; one is "serious" and the other is "not serious." Because the "not serious" culture often mimics the other, we might conclude that it is basically a "school" in charge of training children for adult life. The same idea is found, again and again, in the literature on play behavior not only of humans but also of other animals. Even if it is so, it is an open question what is really and importantly being learned. Maybe even the idea of learning is misleading or limited, when we try to understand the so-called preparation provided by the "school." There are so many possibilities that the acquisition of skills such as those involved in hunting, gathering, cooking, and other specifiable techniques may be a secondary, or even irrelevant, effect of play behavior. Less specifiable and less concrete areas such as social abilities, experiencing and sharing different emotions, feelings, and sensations, understanding values, and practicing an adherence to rules, conventions, and language, are equally possible areas in which learning could be taking place.

Children at play do not look like children in the classroom. They seem to be enjoying themselves and they laugh a lot. Play is clearly a leisurely endeavor. However, the fact that they have fun does not mean that learning is not taking place. There is a lot to be learned regarding the games themselves: their rules and the details of how to behave. What remains to be demonstrated is the extent to which the skills and competences relevant to their future adult life are being learned during play behavior. It may well be that such long-term effects exist; however, at face value, play is just fun. Its role in preparing children for adult life may be subtle and difficult to demonstrate by observation or experiment. Fun and enjoyment are suggestive of fluency and full mastery of the tasks at hand. The repetitive character of many forms of play, which might be taken as drilling, may well be either simple fun or something that goes beyond the technique and allows other subtler psychological experiences to emerge. The obvious redundancy of everyday play behavior may simply mean that there is nothing else to do. Spending the first years of life at play may simply be the easiest way of living, if we accept the idea that there is an innate readiness for culture. The younger children will join the group and eagerly get into the play culture not because they are told to do that but because they feel like playing the game.

Even babies, when playing with their mothers, give clear indications that repetition allows for new experiences. Take, for instance, the situation in which mother and child repeatedly enact a sequence that ends with the child being tickled by the mother, or those in which some fear is felt by the child, but then everything ends in laughter (for urban children: Murray &

Andrews, 2000; Papousek & Papousek, 1984; Seidl de Moura & Ribas, 2000, 2002). Eibl-Eibesfeldt´s (1989) observation of Kosarek and Eipo children, and those in Western New Guinea, Yanomami, Brazilian Indian groups, Himba, G/wi, Eipo, !Kung, and Trobriand, are suggestive that such aspects of repetition may be generalized to other cultures, as can be seen in the vast photographic material collected by him. As such situations are repeated, it becomes clear that the baby anticipates the whole sequence. The element of surprise is no longer present; the sequence becomes like a theater play in which both actors play their roles. Invitations, postural facilitations of the behavior involved, and laughter a little ahead of its proper moment are the signals of the theatrical nature of the repetition.

It is quite possible that the major effects of play relate to such areas as personality development and fluent mastery of culture. When an Indian boy is given his first little toy bow and arrows by an adult and begins to try and shoot with them, the most important consequence may be found in areas other than the development of a skill. His experience of being given the little bow and arrows encompasses the affective relationship with the adult, the emotion of becoming the owner of his toy and doing something that adults do with their big bows and arrows, and overall familiarity with the toy and its way of working. It will be clear to him that he is now a little man. The actual skill of arrow shooting may well be a secondary consequence. Such skill may not result from the shooting exercises in the way that a skill training procedure is intended to achieve. Thus play behavior is not directly involved in turning children into effective hunters, competent mothers, and so on. Such consequences, if real, may be quite indirect.

It may turn out to be difficult to find, demonstrate, and measure the effects of play, if it is true that they are indirect, subtle, cumulative, and multiple in the sense of involving emotions and affects. Smith and Simon (1984), discussing research on the effect of play on subsequent performance, noted that experimental studies in Western societies customarily employed one-session designs: "Some important things may indeed be learnt from a single experience . . . but what we know of naturally occurring play in animals and humans does not suggest that learning in play is of this kind. . . . The probable inference is that any learning from play is piecemeal, and cumulative over bouts and episodes" (p. 212). Smith (2002), reviewing literature on the effects of pretend play on subsequent performance, has pointed out a set of methodological hurdles that exist even in relatively well-defined problems of searching for causal links.

Many forms of play and games are not rehearsals of future adult skills. The best way to describe play and games—and, just as important, the way children deal with them—is to understand that the concept of culture applies. As Carvalho and Pontes (2003) have noticed "plays are like transmitted rituals repeated or re-created. Such transmission takes place within the play group without adult interference, from older to younger children" (pp. 15–16). Additionally, Carvalho and Rubiano (2004), when discussing a new kind of play developed by the children they were observing, remarked that "the idiosyncratic plot, developed by the children and at first shared by a few other individuals, has the potential to become *traditional*, that is, to be shared by the whole group and be transmitted to new members by ontogenetic transmission processes" (p. 186).

The young child enters a culture as he or she joins the group of older children, and he or she will become fluent in that culture just as he or she becomes fluent in language. The child is playing at culture. In addition, these children all know that their culture is a play culture and that the "real thing" is the adult culture, with which they are in full contact by observation and partial involvement. When the time comes for boys and girls to leave, rather abruptly, their play culture and enter adult life, they will be ready to do so, not only because they have been watching it and even, to a certain degree, interacting with it, but because they have exercised *being cultural* for years.

Play, judged according to the criteria of nonliterality, positive affect, flexibility (Krasnor & Pepler, 1980; Smith & Vollstedt, 1985), attention to means rather than ends, and guided by the implicit question "What can I do with this object?", in contrast with the question "What is the object and what can it do?" that is characteristic of exploratory behavior (Hutt, 1966) appears to be ubiquitous among forager children. They exhibit all the forms of play described in children from urban samples: exercise play, play with objects, social contingency play, rough-and-tumble play, and games with rules.

Forager children have a relatively long period of pressure-free immaturity, during which they play in an atmosphere that is free of outside tension. There may be such a phenomenon as an inner tension inherent in the nature of the play itself. However, this is a "play tension," which should not stop us from understanding that the overall setup in which children are playing is a relaxed atmosphere. Playful activity offers an opportunity for a lot of creative activity. Many special skills and behaviors important in adult life can be practiced in a protected environment, before the need to use them in serious contexts arises (Bruner, 1976; Dolhinow & Bishop, 1970; Smith, 1982). In play the consequences of children's actions are minimized, and they can learn in less risky situations. Limits can be tested with relative impunity.

On the whole, there is little intentional pedagogy in hunter-gatherer societies, but children do learn from adults, who serve mainly as models and sources of affection (Bruner, 1965, 1976). To the eyes of observers who have been raised and educated in modern societies, the absence of such things as schools, classes, lectures, and even clear-cut sessions of emphatic teaching may convey the impression that children are not taught by adults at all and that they only learn by observation. It may well be, however, that by means of informal, subtle, and one-trial teaching, adults may be intentionally transmitting much more than we realize because of our concepts of what education is.

Play takes place in a relaxed emotional state. In this relaxed field children defy limits and face dangers. They have to trust peers, because they can be hurt if their partners cheat. In the game "Where is the Fire?", a Camaiurá child accepts being covered with sand and breathing through a tunnel (Moisés, 2003). In a form of play of the Mehináku, which combines physical exercise and social contingency, a child allows a rope to be put around his or her neck (Gregor, 1982). The peer must have made the false knot correctly, so that it will undo when the ends are suddenly pulled. We use the expression "a knot in the throat" to refer to an uncomfortable feeling caused by a big problem. The Mehináku act it out. Tension rises and is suddenly relieved. Something frightening suddenly turns out into nothing. This is the essence of humor.

Difficulties are added to play in order to show off intelligence and physical skill. The mud of the bottom of a lake is shuffled by the Camaiurá children to darken the water and impede the vision of the chasing partner. They also play rough and tumble in pairs in the water, one child standing on the shoulders of the other (Moisés, 2003). Standing on an unstable base and running the risk of falling down is another cue, in addition to laughter and smiling, that the contest is not serious.

Qualities such as strength, courage, and resistance to pain are exhibited and reaffirmed during play, mainly among males. For instance, pairs of Xavante boys, their faces painted with the colors of different clans, beat on each others' arms with roots (Bastos, 2001). This seems to be a culturally ritualized form of play that falls halfway between rough-and-tumble play and a game with rules (cf. Fry, Chapter 4, this volume).

Showing off superior physical skills and intelligence coexists with self-handicapping when older or stronger children interact with younger or weaker peers.

Pretend play, too, appears to be ubiquitous in forager societies. According to Harris (1989), "the stable timing of its onset in different cultures strongly suggests a neuropsychological timetable and a biological basis" (p. 256). Among the Parakanã, *Konomia* (4- to 6-year-olds) exhibit more pretending than *Otyaro* (7- to 12-year-olds). This finding is in accordance with the inverted-U developmental curve reported for urban samples (cf. Smith, Chapter 8, this volume). Elaborated reports of make-believe, such as *teneju itãi* and *yetamá* among the Mehináku, and *tekatawa* among the Parakanã, contradict the impression of "impoverished play" in hunter-gatherers. These are complex episodes involving scripts and roles, such as marriage, birth, sickness, and death.

We agree with Slaughter and Dombrowski (1989) that "children's social and pretend play appears to be biologically based, sustained as an evolutionary contribution to human psychological growth and development. Cultural factors regulate the amount and type of expression of these play forms" (p. 290). According to Bruner (1976), "play can serve as a vehicle for teaching the nature of a society's conventions and it can also teach about the nature of convention per se" (p. 49). The promotion of fluency with cultural rules and conventions may be a main function of play.

Variations in time spent playing by foragers have been reported (Blurton Jones, 1993, Blurton Jones & Konner, 1973; Konner, 1972) as a function of costs and benefits involved in foraging. !Kung children have a longer period of childhood, free from hunting and gathering, and spend more time in camp playing with peers than Hadza children. In the case of the !Kung, hunting and gathering is relatively dangerous, whereas in the case of the Hadza, children can safely and efficiently engage in productive activities.

Blurton Jones (1993) reported that gender differences characterize the time when childhood idleness of the !Kung comes to an end. Whereas young women marry around the age of 15, men do not marry until their late 20s. Serious engagement in gathering begins earlier than hunting. The expression "owners of the shade" is used to refer to the younger men still enjoying their prescribed idleness before learning about hunting by accompanying older relatives on hunts.

Gender variations in time spent playing have also been found among Parakanã Indians by Gosso. In the *Konomia* age class (4–6 years old) there are no differences between the sexes, but at the *Otyaro* age class (7–12 years old), boys spend more time playing then girls. Girls are engaged in productive activities earlier than boys.

Forager societies such as those still found in Brazil are privileged settings for studies of play behavior. In spite of so-called contact with civilization and the practice of incipient agriculture, many of their cultural features can be taken as representative of what happened during human evolution. We do not know how accurate a picture of the past such societies offer us, but it is certainly much less inaccurate than that we can find in modern societies. It seems worthwhile to try and overcome the enormous practical difficulties involved in gaining access to them in order to carry out studies of play behavior.

ACKNOWLEDGMENTS

A grant of Conselho Nacional de Desenvolvimento Científico e Tecnológico supported this work. We thank Olga Cepeda for her assistance with the translation of the first version of the manuscript, José Sávio Leopoldi for valuable suggestions of various anthropological studies on Indian play, and the editors for valuable suggestions and insightful comments. We are also grateful to Laura Patchkofsky for her help with the manuscript.

NOTES

1. The !Kung speech is one of the most phonetically complex languages. It is composed of many click sounds. In the sound "≠" the front part of the tongue is pressed against the alveolar ridge. On release a sharp, flat snap is produced (Wannenburgh, 1979).
2. Effect of age class on exercise play (analysis of variance [ANOVA]): $F_{1,25} = 1.57, p = .22$; effect of sex: $F_{1,25} = 1.64, p = .21$; interaction effect, age class × sex: $F_{1,25} = 0.28, p = .60$.
3. The regions mentioned by Cooper were Middle American, Antillean, Orinocoan, Amazonian, Guianan, Eastern Brazilian, Eastern Bolivian, and Chacoan (*Chané*).
4. Effect of age class on construction play (ANOVA): $F_{1,25} = 2.27, p = .14$; effect of sex: $F_{1,25} = 0.54, p = .47$; interaction effect, age class × sex: $F_{1,25} = 12.15, p < .01$.
5. Cooper refers specifically to people known as Ona, Araucanian, Guarani, and Mosquito-Sumo.
6. Effect of age class on social contingency play (ANOVA): $F_{1,25} = 0.29, p = .59$; effect of sex: $F_{1,25} = 3.97, p = .057$; interaction effect, age class × sex: $F_{1,25} = 0.92, p = .35$.
7. Effect of age class on rough-and-tumble play (ANOVA): $F_{1,25} = 0.02, p = .88$; effect of sex: $F_{1,25} = 1.03, p = .32$; interaction effect, age class × sex: $F_{1,25} = 0.37, p = .55$.
8. The regions mentioned by Cooper are Central American, Antillean, Andean, Araucanian, Orinocoan, Amazonian, Guianan, Eastern Brazilian, Eastern Bolivian, Chacoan, Pampean, Patagonian, and Fuegian.
9. Effect of age class on fantasy play (ANOVA): $F_{1,25} = 6.80, p < .01$; effect of sex: $F_{1,25} = 2.57, p = .12$; interaction effect, age class × sex: $F_{1,25} = 0.10, p = .75$.
10. Effect of age class on games with rules (ANOVA): $F_{1,25} = 12.89, p < .001$; effect of sex: $F_{1,25} = 3.54, p = .07$; interaction effect, age class × sex: $F_{1,25} = 0.70, p = .78$.

REFERENCES

Bastos, K. (2001). *Brincadeira de criança: Como os pequenos índios se divertem* [Child play: How indigenous children have fun]. *Brasil Indígena—Fundação Nacional do Índio, 1*, 25–27.

Bichara, I. D. (1999). *Brincadeira e cultura: o faz-de-conta das crianças Xocó e do Mocambo (Porto da Folha/SE)* [Play and culture: Fantasy play of the Xocó and Mocambo children]. *Temas em Psicologia, 7*, 57–64.

Bichara, I. D. (2003). *Nas águas do Velho Chico* [In the waters of San Francisco River]. In A. M. A. Carvalho, C. M. C. Magalhães, F. A. R. Pontes, & I. D. Bichara (Eds.), *Brincadeira e cultura: Viajando pelo Brasil que brinca* [*Play and culture: Traveling through Brazil which plays*] (pp. 89–107). São Paulo, Brazil: Casa do Psicólogo.

Bjorklund, D. F. (1997). The role of immaturity in human development. *Psychological Bulletin, 122*, 153–169.

Bjorklund, D. F., & Pellegrini, A. D. (2002). *The origins of human nature: Evolutionary developmental psychology.* Washington, DC: American Psychological Association.

Blurton Jones, N. (1993). The lives of hunter-gatherer children: Effects of parental behavior and parental reproductive strategy. In M. E. Pereira & L. A. Fairbanks (Eds.), *Juvenile primates: Life history, development, and behavior* (pp. 309–325). New York: Oxford University Press.

Blurton Jones, N., & Konner, M. (1973). Sex differences in behaviours of London and Bushman children. In R. P. Michael & J. H. Crook (Eds.), *Comparative ecology and behavior of primates* (pp. 690–750). New York: Academic Press.

Boulton, M. J. (1992). Participation in playground activities in middle school. *Educational Research, 34*, 167–182.

Bruner, J. S. (1965). The growth of mind. *American Psychologist, 20*, 1007–1017.

Bruner, J. S. (1976). Nature and uses of immaturity. In J. S. Bruner, A. Jolly, & K. Sylva (Eds.), *Play: Its role in development and evolution* (pp. 28–64). New York: Penguin Books.

Buss, D. M. (2000). The evolution of happiness. *American Psychologist, 55*, 15–23.

Bussab, V. S. R., & Ribeiro, F. L. (1998). *Biologicamente cultural* [Biologically cultural]. In L. Souza, M. F. Q. Freitas, & M. M. P. Rodrigues (Eds.), *Psicologia: Reflexões (im)pertinentes* [*Psychology: (Im)pertinent reflections*] (pp. 175–193). São Paulo, Brazil: Casa do Psicólogo.

Byers, J. A. (1984). Play in ungulates. In P. K. Smith (Ed.), *Play in animals and humans* (pp. 43–65). Oxford, UK: Blackwell.

Caldera, Y. M., O'Brien, M., Truglio, R. T., Alvarez, M., & Huston, A. C. (1999). Children's play preferences, construction play with blocks, and visual–spatial skills: Are they related? *International Journal of Behavioral Development, 23*, 855–872.

Carvalho, A. M. A., & Pontes, F. A. R. (2003). *Brincadeira é cultura* [Play is culture]. In A. M. A. Carvalho, C. M. C. Magalhães, F. A. R. Pontes, & I. D. Bichara (Eds.), *Brincadeira e cultura: Viajando pelo Brasil que brinca* [*Play and culture: Traveling through Brazil which plays*] (pp. 15–32). São Paulo, Brazil: Casa do Psicólogo.

Carvalho, A. M. A., & Rubiano, M. R. B. (2004). *Vínculo e compartilhamento na brincadeira de criança* [Attachment and sharing in children's play]. In M. C. Rossetti-Ferreira, K. de S. Amorim, A. P. S. da Silva, & A. M. A. Carvalho (Eds.), *Rede de significações e o estudo do desenvolvimento humano* [*Network of meanings and human development study*] (pp. 171–187). Porto Alegre, Brazil: Artmed.

Clark, C. B. (1977). A preliminary report on weaning among chimpanzees of the Gombe National Park, Tanzania. In S. Chevalier-Skolnikoff & F. E. Poirier (Eds.), *Primate bio-social development: Biological, social, and ecological determinants* (pp. 235–260). New York: Garland Publishing.

Clastres, P. (1988). *A sociedade contra o Estado: Pesquisa de antropologia política* [*A society against the State: Investigation in political anthropology*]. São Paulo, Brazil: Cosac & Naify.

Cohn, C. (2002). *A experiência da infância e o aprendizado entre os Xikrin* [Childhood experiences and learning among the Xikrin]. In A. L. Silva, A. V. L. S. Macedo, & A. Nunes (Eds.), *Crianças indígenas: Ensaios antropológicos* [*Indigenous children: Anthropological essays*] (pp. 117–149). São Paulo, Brazil: Global.

Cooper, J. M. (1949). Games and gambling. In J. H. Steward (Ed.), *Handbook of South American Indians: The comparative ethnology of South American Indians* (pp. 503–524). Washington, DC: U.S. Government Printing Office.

Cullumbine, H. (1950). Heat production and energy requirements of tropical people. *Journal of Applied Physiology, 2*, 201–210.

DeVore, I., & Konner, M. J. (1970). Infancy in hunter-gatherer life: An ethological perspective. In N. F. White (Ed.), *Ethology and psychiatry* (pp. 113–141). Toronto: University of Toronto Press.

DiPietro, J. A. (1981). Rough and tumble play: A function of gender. *Developmental Psychology, 17*, 50–58.

Dolhinow, P. J., & Bishop, N. (1970). The development of motor skills and social relationships among primates through play. In J. P. Hill (Ed.), *Minnesota Symposium of Child Psychology* (pp. 180–198). Minneapolis: University of Minnesota Press.

Draper, P. (1976). Social and economic constraints on child life among the !Kung. In R. B. Lee & I. DeVore (Eds.), *Kalahari hunter-gatherers: Studies of the !Kung San and their neighbors* (pp. 199–217). Cambridge, MA: Harvard University Press.

Draper, P., & Cashdan, E. (1988). Technological change and child behavior among the !Kung. *Ethnology, 27*, 339–365.

Eaton, W. O., & Yu, A. P. (1989). Are sex differences in child motor activity level a function of sex differences in maturational status? *Child Development, 60*, 1005–1011.

Eibl-Eibesfeldt, I. (1974). The myth of the aggression-free hunter and gatherer society. In R. L. Holloway (Ed.), *Primate aggression, territoriality and xenophobia* (pp. 435–457). London: Academic Press.

Eibl-Eibesfeldt, I. (1989). *Human ethology.* New York: Aldine de Gruyter.

Fein, G. G. (1981). Pretend play in childhood: An integrative review. *Child Development, 52*, 1095–1118.

Gaskins, S. (2000). Children's daily activities in a Mayan Village: A culturally grounded description. *Cross-Cultural Research, 34*, 375–389.

Geary, D. C., & Bjorklund, D. F. (2000). Evolutionary developmental psychology. *Child Development, 71*, 57–65.

Gosso, Y., Morais, M. L. S., & Otta, E. (2002, October). *Estudo comparativo do faz-de-conta de crianças de 3 comunidades brasileiras: um grande centro urbano, uma zona rural e uma aldeia indígena* [Comparative study of pretend play in three Brazilian communities: A large city, a rural village and an Indian village]. Paper presented at the 42nd Annual Meeting of Psychology, Florianópolis, Brazil.

Gosso, Y., & Otta, E. (2003). *Em uma aldeia Parakanã* [At a Parakanã village]. In A. M. A. Carvalho, C. M. C. Magalhães, F. A. R. Pontes, & I. D. Bichara (Eds.), *Brincadeira é cultura: Viajando pelo Brasil que brinca* [*Play and culture: Traveling through Brazil which plays*] (Vol. 1, pp. 33–76). São Paulo, Brazil: Casa do Psicólogo.

Gregor, T. (1982). *Mehináku: O drama da vida diária em uma aldeia do Alto Xingu* [*Mehináku: The daily drama of a village of the Alto Xingu*]. São Paulo, Brazil: Nacional.

Harris, P. (1989). *Children and emotion: The development of psychological understanding.* Oxford, UK: Blackwell.

Hewlett, B. S., Shannon, D., Lamb, M. E., Leyendecker, B., & Schölmerich, A. (1998). Culture and early infancy among central African foragers and farmers. *Developmental Psychology, 34*, 653–661.

Humphreys, A. P., & Smith, P. K. (1987). Rough-and-tumble play, friendship, and dominance in school children: Evidence for continuity and change with age. *Child Development, 58*, 201–212.

Hutt, C. (1966). Exploration and play in children. *Symposia of the Zoological Society of London, 18*, 61–81.

Hutt, C. (1972). Sex differences in human development. *Human Development, 15*, 153–170.

Konner, M. (1972). Aspects of the developmental ethology of a foraging people. In N. Blurton Jones (Ed.), *Ethological studies of child behaviour* (pp. 285–304). Cambridge, UK: Cambridge University Press.

Konner, M. (1977). Infancy among the Kalahari desert San. In P. H. Leiderman, S. R. Tulkin, & A. Rosenfeld (Eds.), *Culture and infancy: Variations in the human experience* (pp. 287–301). New York: Academic Press.

Krasnor, L. R., & Pepler, B. J. (1980). The study of children's play: Some suggested future directions. In K. Rubin (Ed.), *Children's play* (pp. 85–95). San Francisco: Jossey-Bass.

Leacock, E. (1976). At play in African villages. In J. S. Bruner, A. Jolly, & K. Sylva (Eds.), *Play: Its role in development and evolution* (pp. 466–464). London: Penguin Books.

Leakey, R. E. (1981). *A evolução da humanidade [The evolution of humankind].* São Paulo, Brazil: Melhoramentos.

Lordelo, E. R., & Carvalho, A. M. A. (1989). *Comportamento de cuidado entre crianças:*

Uma revisão [Caretaking among children: A review]. *Psicologia: Teoria e Pesquisa, 5*, 1–19.

Martini, M. (1994). Peer interactions in Polynesia: A view from the Marquesas. In J. L. Roopnarine, J. E. Johnson, & F. H. Hooper (Eds.), *Children's play in diverse cultures* (pp. 73–103), Albany: State University of New York Press.

McGrew, W. C. (1972). *An ethological study of children's behaviour.* London: Academic Press.

Mead, M. (1949). *Coming of age in Samoa: A psychological study of primitive youth for Western civilization.* New York: New York American Library.

Moisés, D. (2003, December 15). *Expedição conta como brincam os índios: Brinquedos, jogos e brincadeiras* [An expedition tells how Indians play: Toys, games and tricks]. *O Estado de São Paulo*, p. A8.

Morais, M. L. S., & Carvalho, A. M. A. (1994). *Faz-de-conta: Temas, papéis e regras na brincadeira de crianças de quatro anos* [Fantasy play: Themes, roles, and rules in preschool children play]. *Boletim de Psicologia, 100/101*, 21–30.

Morais, M. L. S., & Otta, E. (2003). *Entre a serra e o mar* [In between the mountains and the sea]. In A. M. A. Carvalho, C. M. C. Magalhães, F. A. R. Pontes, & I. D. Bichara (Eds.), *Brincadeira é cultura: Viajando pelo Brasil que brinca [Play and culture: Traveling about in Brazil which plays]* (pp. 127–156). São Paulo, Brazil: Casa do Psicólogo.

Murray, L., & Andrews, L. (2000). *The social baby.* Richmond, VA: CP Publishing.

Nunes, A. (1999). *A sociedade das crianças A'uwe-Xavante: Por uma antropologia da criança [The society of the A'uwe-Xavante children: For children's anthropology]*. Lisboa, Portugal: Ministério da Educação/Instituto de Inovação Cultural.

Nunes, A. (2002). *No tempo e no espaço: Brincadeiras das crianças A'uwe-Xavante* [In time and space: Games of the A'uwe-Xavante children]. In A. L. Silva, A. V. L. S. Macedo, & A. Nunes (Eds.), *Crianças indígenas: Ensaios antropológicos [Indigenous children: Anthropological essays]* (pp. 64–99). São Paulo, Brazil: Global.

Papousek, H., & Papousek, M. (1984). Learning and cognition in the everyday life of human infants. *Advances in the Study of Behavior, 14*, 127–159.

Parker, S. T. (1984). Playing for keeps: An evolutionary perspective on human games. In P. K. Smith (Ed.), *Play in animals and humans* (pp. 271–293). New York: Blackwell.

Pellegrini, A. D. (1984). Identifying causal elements in the thematic fantasy play paradigm. *American Education Research Journal, 19*, 443–452.

Pellegrini, A. D. (1989). Elementary school children's rough-and-tumble play. *Early Childhood Research Quarterly, 4*, 245–260.

Pellegrini, A. D., Horvat, M., & Huberty, P. D. (1998). The relative costs of children's physical activity play. *Animal Behaviour, 55*, 1053–1061.

Pellegrini, A. D., & Smith, P. K. (1998a). The development of play during childhood: Forms and possible functions. *Child Psychology and Psychiatry Review, 3*, 51–57.

Pellegrini, A. D., & Smith, P. K. (1998b). Physical activity play: The nature and function of a neglected aspect of play. *Child Development, 69*, 577–598.

Peller, L. E. (1971). Models of children's play. In R. E. Herron & B. Sutton-Smith (Eds.), *Child's play* (pp. 110–125). New York: Wiley.

Pereira, M. C. S. (1998). *Meninas e meninos Kaingáng: O processo de socialização [Kaingáng girls and boys: The socialization process]*. Londrina, Brazil: Editora Universidade Estadual de Londrina.

Piaget, J. (1962). *Play, dreams, and imitation in childhood*. New York: Norton.

Piaget, J. (1965). *The moral judgement of the child*. New York: Free Press.

Ribeiro, B. G. (1987). *Suma etnológica Brasileira: Etnobiologia [Brazilian ethnological studies: Ethnobiology]*. Petrópolis, Brazil: Vozes.

Routh, D., Schroeder, C., & O'Tuama, L. (1974). Development of activity levels in children. *Developmental Psychology, 10*, 163–168.

Seidl de Moura, M. L., & Ribas, A. F. P. (2000). *Desenvolvimento e contexto sociocultural: A gênese da atividade mediada nas interações iniciais mãe-bebê* [Development and sociocultural contexts: The genesis of the mediated activity in the initial mother–baby interactions]. *Psicologia: Reflexão e Crítica, 13*, 245–256.

Seidl de Moura, M. L., & Ribas, A. F. P. (2002). *Imitação e desenvolvimento inicial: Evidências empíricas, explicações e implicações teóricas* [Imitation and initial development: Empirical evidences, explanations, and theoretical implications]. *Estudos de Psicologia, 7*, 207–215.

Shostak, M. (1976). A !Kung woman's memories of childhood. In R. B. Lee & I. DeVore (Eds.), *Kalahari hunter-gatherers: Studies of the !Kung San and their neighbors* (pp. 246–277). Cambridge, MA: Harvard University Press.

Shostak, M. (1981). *Nisa: The life and words of a !Kung woman*. Harmondsworth, UK: Penguin Books.

Sinker, M., Brodin, J., Fagundes, V., Kim, F., Hellberg, G., Lindberg, M., Trieschmann, M., & Björck-Akensson, E. (1993). *Children's concept of play: A study in four countries*. Stockholm, Sweden: Samhall Kalmarsund.

Slaughter, D., & Dombrowski, J. (1989). Cultural continuities and discontinuities: Impact on social and pretend play. In M. N. Block & A. D. Pellegrini (Eds.), *The ecological context of children's play* (pp. 282–310). Norwood, NJ: Ablex.

Smith, P. K. (1982). Does play matter? Functional and evolutionary aspects of animal and human play. *Behavioral and Brain Sciences, 5*, 139–184.

Smith, P. K. (2002). Pretend play, metarepresentation and theory of mind. In R. Mitchell (Ed.), *Pretending and imagination in animals and children* (pp. 129–141). Cambridge, UK: Cambridge University Press.

Smith, P. K., & Connolly, K. (1980). *The ecology of preschool behaviour.* Cambridge, UK: Cambridge University Press.

Smith, P. K., & Simon, T. (1984). Object play, problem-solving and creativity in children. In P. K. Smith (Ed.), *Play in animals and humans* (pp. 199–216). New York: Blackwell.

Smith, P. K., & Vollstedt, R. (1985). On defining play: An empirical study of the relationship between play and various play criteria. *Child Development, 56*, 1042–1050.

Sutton-Smith, B., & Roberts, J. M. (1981). Play, toys, games, and sports. In H. C.

Triandis & A. Heron (Eds.), *Handbook of cross-cultural psychology: Developmental psychology* (Vol. 4, pp. 425–471). Boston: Allyn & Bacon.

Sutton-Smith, B., Rosenberg, B. G., & Morgan, E. F., Jr. (1963). Development of sex differences in play choices during adolescence. *Child Development, 34*, 119–126.

Truswell, A. S., & Hansen, J. D. L. (1976). Medical research among the !Kung. In R. B. Lee & I. DeVore (Eds.), *Kalahari hunter-gatherers: Studies of the !Kung San and their neighbors* (pp. 166–195). Cambridge, MA: Harvard University Press.

Turnbull, C. (1961). *The forest people.* New York: Simon & Schuster.

Wannenburgh, A. (1979). *The bushman.* Singapore: Tien Wah Press.

CHAPTER 10

Farming, Foraging, and Children's Play in the Okavango Delta, Botswana

JOHN BOCK

Few studies have examined play in either foraging or traditional agrarian contexts. Konner (1972) studied play among the Ju'/hoansi of Botswana at a time when a large proportion of the population was solely dependent on foraging. He found that all elements of play normally considered by developmental psychologists were present and that the children's play involved elaborate representations of the socioecological context in which they lived (see Sutton-Smith, 1971), a finding mirrored by a subsequent comparison of Ju'/hoansi and English children (Blurton Jones & Konner, 1973). In addition to my own work in the Okavango Delta, there have been three other studies of children's play in foraging societies. Kamei (in press) has studied play among Baka children in Cameroon and found that substantial time is spent in pretend play that reflects the socioecological factors affecting children's development in that society. Blurton Jones (1993) observed Hadza children in Tanzania making toys that were used in pretend play that reflected the activities and events integral to the children's lives. In her study of Canadian Inuit children's play, Briggs (1990) found that children engaged in "playwork" that was a vehicle for the acquisition of problem-solving and survival skills appropriate to the harsh circumarctic environment.

It has been argued that some of the value of studying children's play in foraging societies is the insight afforded into adaptation to earlier foraging environments, such as those of our Pleistocene ancestors (Pellegrini & Gustafson, Chapter 6, this volume; Smith, in press). Recent research on the evolution of childhood has identified salient features of foraging socio-ecological contexts that may have been selective forces on features of the human juvenile period (Bock & Sellen, 2002). These include the time investment for skill acquisition and the development of physical attributes necessary to become competent foragers (Bock, 2004) as well as the trade-offs between immediate and deferred benefits of children's activities relative to the reproductive and economic interests of children and their parents (Bock, 1995, 2002a, 2002b). By understanding features of foraging environments that acted as selective pressures on the range and form of children's developmental trajectories in contemporary humans, the study of children's activities in varied socioecological contexts should yield insights relevant to the evolutionary history of childhood. However, we should always keep in mind that contemporary foraging societies display tremendous variability. The time allocation of children, the age-dependent productivity of children, and the tradeoff between immediate and deferred benefits differ widely among foraging societies (Bock & Sellen, 2002).

Children's activities, including play, in agrarian societies are a function of the interaction of the same evolved features of childhood and the cost–benefit structure of children's activities in different socioecological contexts. Studies of play in traditional agrarian societies have had a broader focus than those in foraging societies, incorporating issues of interest in the development of theory in psychology, such as the delineation of play stages in the Piagetian sense (Schwartzman, 1976). Several studies of children's play in agrarian societies have focused on the enculturation function and the development of psychosocial and cognitive skills specific to adult roles.

Lancy (1996) found that Kpelle children of Liberia engaged in play activities that were directly related to the development of skills needed to achieve adult competence in Kpelle society. Other research has focused on the difficulty of classifying activities as productive or nonproductive in agrarian economic systems (for a discussion of this issue, see Bock & Johnson, 2004). Katz (1991) found that children among subsistence farmers in southern Sudan combined work and play in ways that allowed them to develop environmental knowledge pertinent both to their lives as children and their future lives as adults. Punch (2003) found that the play of rural Bolivian children was highly constrained by the material and social context. As children's workloads increased with age, they incorporated more play into their work lives. Gosso and Otta (2003) examined children's play in another mixed subsistence economy of foraging and horticulture among Parakanã Amerindians of Amazonian Brazil. They found that pretense and object play were the most common forms of play among young children and that participation in rule-oriented and game forms of play increased with age.

In this chapter, I examine the socioecological context of children's play in a traditional community in the Okavango Delta of Botswana, in order to develop a better understanding of the possible influences of play on children's development of adult competence. The mixed subsistence economy of the study community provides the means to identify the common and divergent elements of agrarian and foraging subsistence regimes' influence on children's play. Although only a small proportion of the people in the present study obtain all their nutritional requirements from foraging, all of them obtain substantial portions of their needs from fishing, hunting, and the collection of wild plant sources. No longer nomadic, and with elements of social structure strongly influenced by both economic change and long social interchange with other groups, the people in the study community contribute another dataset of play behavior among foraging people, and, at the same time, advance our knowledge of play in the context of agrarian socioecology. As a multiethnic community of forager/agropastoralists, this group at once illustrates the flexible adaptation of childhood and the joint effects of multiple socioecological contexts on children's time allocation and behavior.

THE FUNCTIONALIST PERSPECTIVE

Across disciplines, there is agreement that play is generally defined as an activity in which only juveniles engage (for reviews, see Barber, 1991; Bekoff & Byers, 1998; Bjorklund & Pellegrini, 2002; Bloch & Pellegrini, 1989; Fagen, 1981; Pepler & Rubin, 1982; Power, 2000; Smith, 1982). Beyond that, definitions of play are usually group into structural and functional categories, which, according to Martin and Caro (1985), amount "to little more than statements that 'play is playful behavior,' " p. 59). The theoretical perspectives on play across disciplines fall into two camps that I call "functionalists" and "neutralists."

Functionalists argue that play is costly and therefore must have commensurate benefits. If play is energetically costly, it must have proportionately large benefits, at some point, for organisms to have evolved this behavior. Biologists hypothesize that these benefits are realized in three aspects of development: motor training, socialization, and cognitive or sensorimotor training (Bekoff & Byers, 1981; Smith, 1982). Similarly, psychologists and others studying human development hypothesize that play is an important factor in the proper development of such qualities as role taking, quantitative invariance, language acquisition, problem solving, creativity, and divergent thinking (Rubin, Fein, & Vandenberg, 1983). I (Bock, 1995, 2002b; Bock & Johnson, 2004) have emphasized the functional quality of play in the development of skills applicable to productive tasks.

The three main forms of children's play commonly characterized by researchers—object, physical (exercise, rough and tumble [R&T]), and fantasy (or pretend) (Cohen, 1987)—have been recognized, to some extent, in all human societies (Eibl-Eibesfeldt, 1989; Roopnarine, Johnson, & Hooper, 1994; Schwartzman, 1978; Slaughter & Dombrowski, 1989). If children's play has a functional quality, then the frequency of each type of play should vary in accordance with social and ecological features of a given society (Bock, 2002b; Bock & Johnson, 2004).

For instance, object play has been hypothesized as helping children to become familiar with tools, materials, and/or processes (Bock, 1995, 2002b; Bock & Johnson, 2004; Geary, 2002; Pellegrini & Bjorklund, 2004; Smith & Simon, 1984) or developing their skills through motor training (Bock, 1995, 2002b; Bock & Johnson, 2004; Byers & Walker, 1995). If this were the case, object play should appear in forms that echo adult productive tasks in a given society or serve to enhance performance in precursor or component tasks (Bock & Johnson, 2004). The frequency of object play in a given society would be expected to reflect the age-specific learning curve of adult object-oriented tasks in that subsistence ecology (Bock, 2002b). Traditional hunting, for example, has been hypothesized to be a skill-intensive activity with a very long learning curve (Bock, 1995, 1998, 2002b; Hill & Hurtado, 1996; Kaplan & Bock, 2001; Kaplan, Lancaster, Hill, & Hurtado, 2000; Liebenberg, 1990; Walker, Hill, Kaplan, & McMillan, 2002). If this were the case, then we would expect the time allocated to object play by children who will be hunting as adults to be relatively high (Bock & Johnson, 2004). In societies where men are the primary hunters, then, children's object play that is related to the development of hunting ability should be more frequent in boys (Bock, 1995; Bock & Johnson, 2004; Geary, 2002; Pellegrini & Bjorklund, 2004).

R&T play has been hypothesized to function as practice of fighting skills (Fry, 1990; Pellegrini, 2002; Smith, 1982) and the experience of hunting-related aggression (Boulton & Smith, 1992). The frequency of R&T play should then be related to the level of conspecific and interspecific aggression, respectively, in a given socioecological context (see Fry, Chapter 4, this volume). Gender differences in the frequency of R&T play would be expected to be related to sexual selection for aggressiveness, which, in turn, would be directly related to intrasexual competition or hunting acumen (Geary, 2002). Additionally, play fighting and forms of R&T play may be more prevalent in boys in traditional subsistence economies due to the possible deferred benefits to future hunting returns (Geary, 2002; Pellegrini & Smith, 1998; Smith, 1982).

Pretend play has been viewed as preparatory for adult competency because of its rehearsal and role-playing aspects (Bock, 1995; Bock & Johnson, 2004; Konner, 1972; Lancy, 1996; Pellegrini & Bjorklund, 2004; Smith, 1982). This perspective leads to the expectation that the form of pretend play should be related to adult tasks in a given socioecological context. This relation may be due to the high cost of direct practice, or learning by doing, in certain situations (Smith, 1982). For instance, if the use of materials would result in their being lost or spoiled, then pretend play might serve a training function at a lower cost (Bock, 2002a). In addition, we would expect pretend or fantasy play to be highly gender specific, with each gender engaging in that form of play that complements adult roles in a given society (Bock, 1995, 2002b; Bock & Johnson, 2004; Geary, 2002).

THE NEUTRALIST PERSPECTIVE

A contrasting viewpoint, present among both biologists and psychologists, contends that play has little or no cost and little or no benefit. This is what I refer to as the neutralist perspective, because play is seen to be merely an artifact of the development of the neural and musculoskeletal systems, analogous to the epiphenomenal perspective of Gould and Lewontin (1979) in regard to adaptations. Martin and Caro (1985) argue that no studies have demonstrated any benefit to play that could not be gotten through other means with equal cost. There is a consistent theme in human development studies that play may just be a means of using up surplus energy by "blowing off steam" (Rubin, 1982).

Implicit in this neutralist philosophy is the assumption that there is no opportunity cost and that there is an energetic surplus. That is, children could be doing nothing else with their time that would bring a benefit, even if play is not energetically costly, in and of itself. This perspective has persisted because play has been studied in isolation, rather than as a subset of activities within the suite of productive and nonproductive activities in which children can engage.

A LIFE HISTORY PERSPECTIVE ON PLAY

Recent research on the evolution of the primate and human life history pattern has parallels with the alternative perspectives of the play functionalists and neutralists (for reviews, see Bock & Sellen, 2002; Leigh, 2001).

Charnov's (1993; Charnov & Berrigan, 1993) adult mortality model directs attention to low adult mortality as the primary selective force in the evolution of extended juvenility in primates. Low mortality provides additional time for organisms to grow to larger body size before reproducing, thus benefiting from greater energy stores and strength. Hawkes, Blurton Jones, and others have extended this model to humans, focusing on body size rather than cognitive abilities as the constraining factor in the task performance of children relative to adults (Bird & Bliege Bird, 2002; Bliege Bird & Bird, 2002; Blurton Jones, Hawkes, & O'Connell, 1999; Blurton Jones & Marlowe, 2002; Hawkes, O'Connell, Blurton Jones, Alvarez, & Charnov, 1998). One interpretation of the Charnovian model is that cognitive abilities acquired through experiences such as social learning were beneficial, but those benefits were not sufficient selection pressure to extend the human juvenile period (Blurton Jones & Marlowe, 2002).

There are two other models of the evolution of juvenility in primates that are relevant to the role of experience and learning in human development, and that emphasize social learning as an important selective force. The brain-growth model (Bogin, 1999) and the embodied capital model (Kaplan, 1996; Kaplan & Bock, 2001; Kaplan, Lancaster, Bock, & Johnson, 1995; Kaplan et al., 2000) both argue that competence in social and ecological spheres is dependent on skills acquired through experience, especially social learning. These models propose that the benefits of acquiring these skills over a longer childhood outweigh the costs of growing more slowly, resulting in natural selection for a long childhood. I (Bock, 1995, 2002b; Bock & Johnson, 2004) have emphasized the tradeoff between immediate benefits in terms of productivity and deferred benefits in the forms of skills acquired that will be advantageous later in life. The embodied capital model further argues that because the costs of extended juvenility were high, there was additional selection for a complex pattern of intergenerational exchange and lateral food sharing, extensive male parental investment, and extended life span. In addition, the embodied capital model separates the effects of body mass from other factors influencing productivity.

The punctuated development model (Bock, 2002b, 2004; Bock & Johnson, 2002a) continues along the line of reasoning used in the embodied capital model but distinguishes growth-based from experience-based forms of embodied capital. Growth-based forms of embodied capital are attributes such as body size, strength, balance, and general coordination. Experience-based forms of embodied capital are attributes such as cognitive function, motor and muscle memory, memory function, task-specific skills, learned knowledge, endurance, and specific coordination. Growth-based forms tend to be more related to general competency, whereas experience-based forms tend to be more related to specific competency. The performance of any task requires the interaction of growth-based and experience-based embodied capital; the precise mix is a function of task complexity and the physical demands of a given task—and those, in turn, are a function of subsistence ecology.

The punctuated development model proposes that interaction between growth-based and experience-based embodied capital was the primary selective force on the evolution of childhood. We would expect an individual to allocate resources to acquiring growth-based embodied capital when there are experience-based skills that can be applied to utilize that growth, and we would expect an individual to allocate resources to experience-based embodied capital when there is sufficient growth-based embodied capital to apply that experience. In this way, growth- and experience-based embodied capital interact to produce the overall pattern of growth and development, including the sensitive periods.

This perspective can be used to frame the effects of parental investment of time and resources in child competency in terms of the different forms of embodied capital outlined above. In different subsistence ecologies and across different tasks, the amount of investment in growth that would bring a return in learning will vary (Bock, 2002b; Bock & Johnson, 2002a). The degree to which growth constrains learning will vary as a function not only of subsistence ecology but also of the economics of production and may be strongly influenced by the value of labor and the opportunity cost to alternative activities. In foraging economies, as in all others, the variety of tasks performed can be expected to reflect a number of different levels of growth constraints on payoffs to learning. Investment of resources and time by parents in their offspring can be used to build growth-based embodied capital, or it can be used to develop experience-based embodied capital. The optimal solution to the tradeoff between investment in growth- and experience-based embodied capital is expected to be dependent on the societal-based gender- and age-patterning of production, on the specific labor needs of the household, and on the reproductive interests of parents.

HYPOTHESES AND PREDICTIONS

The punctuated development model asserts that the time spent playing should be subject to tradeoffs between immediate productivity and future returns. Moreover, the punctuated development model emphasizes that adult competence is specific to a subsistence regime, and that parental influences on children's time allocation to play should mirror the parents' interests in maximizing returns on children's time allocation, in general (Bock, 2002a, 2002b).

The socioecological perspective guiding this research would predict that the costs and benefits of play can be interpreted by a parent in terms of a tradeoff between immediate productivity and the opportunity costs to immediate productivity and other immediate and/or delayed benefits. For example, if very young children have no measurable productivity within the constraints of a particular subsistence ecology, we should expect that the tradeoff between immediate productivity and future returns for very young children will be strongly in favor of future benefits. If there were a function to play, in terms of general or specific forms of neural, social, and musculoskeletal development, then we would expect parents to want their very young children to play based on some algorithm relating the relationship of those forms of development to the development of productive activity. Of course, it is not surprising that young children do not work very much relative to their play.

When a parent is faced with these allocation decisions across a number of offspring, determining the optimal solution quickly becomes a complicated endeavor. In one-child families, assessing the costs and benefits of investment in different forms of embodied capital is relatively clear-cut, from a theoretical standpoint. With each additional child, this assessment becomes more complex with the addition of opportunity costs and multiple time frames. Parents' reproductive interests are unlikely to coincide with those of any one of their children (Blurton Jones, 1993; Bock, 1995, 1999, 2002a, 2002b; Worthman, 2000). An evolutionary perspective leads us to believe that parents should be concerned with their reproductive interests, as manifest in all of their children. They should even be willing to act to the detriment of a child if doing so benefits the parents themselves. Such conflicts are rife when investment in one child affects investment in others (Trivers, 1972).

This line of reasoning is important, however, in our understanding of children's time allocation to seemingly nonproductive activities such as play. If we now extend the example to children who do have some returns to productive activities, we can make a general qualitative prediction concerning the allocation of time to productive and nonproductive activities from the perspective of the parents. The mix of productive and nonproductive activities should reflect the point where the marginal increase from one more unit of time spent in productive activities would be less than the long-term payoff to the parent of spending that unit of time in some activity related to the development of general or specific skills and strength. In regard to overall time allocation, this payoff structure makes the general prediction that work and play should be negatively related throughout childhood. For a specific activity, play related to that activity should be negatively related to the return rate to that activity, which, in turn, is positively related to the time allocation to that activity (Bock, 2002a, 2002b; Bock & Johnson, 2002a, 2004). Therefore, if we were able to isolate play activities that are directly related to the acquisition of skills toward the performance of an activity, we should see an inverse relationship between the amount of time spent in the play activity and the time spent in performance of the activity. Since we hypothesize that these are due to age-related changes in development, we can further predict that these relationships are a function of age.

To test these predictions, we need to examine the overall pattern of work and play in children of each gender in light of what we know about productive labor. Second, we need to identify specific types of play that are directly related to specific types of productive activities and see if the predicted relationships exist. Unfortunately, there are several methodological issues that make these much easier tasks in the abstract than in reality. What are the different types of nonproductive tasks? Productive activities are generally easy to differentiate, although rest time, embedded activities, or distractions can make that typology more difficult. The differentiation of play activities requires both a typology and the means to accurately place observations into that typology. In the following sections I apply this theoretical perspective to a community with a traditional subsistence ecology and interpret these results in both theoretical and empirical contexts.

METHODS

The Study Community

These issues are examined using data collected in a multiethnic community of approximately 400 people in the Okavango Delta of northwestern Botswana (for detailed description of the study community, see Bock, 1995,

1998; Bock & Johnson, 2002a, 2002b). Five ethnic groups are represented: Bugakhwe, ||Anikhwe, Hambukushu, Dxeriku, and Wayeyi. ||Anikhwe and Bugakhwe people are San speakers who inhabit the Okavango drainage in Namibia and Botswana. ||Anikhwe historically have had a riverine orientation in their foraging, whereas Bugakhwe have been savanna foragers. The ||Anikhwe living in the study community practice a mixed economy, but farm at a much less intensive level than the Bantus. All ||Anikhwe families acquire the bulk of their resources from foraged foods. Moreover, among 50 ||Anikhwe there are only four head of cattle, whereas a typical Bantu homestead of 20 people has an average of 12 head. Bugakhwe in this community are largely oriented toward fishing, hunting, and the collection of wild plant foods. None owns cattle, and a few have small gardens where they grow tobacco and specialty foods such as vegetables.

Hambukushu, Dxeriku, and Wayeyi people are Bantus who inhabit the Okavango River drainage from Angola through the Caprivi Strip of Namibia into northern Botswana. Historically, they have participated in mixed economies of farming; fishing, hunting, and the collection of wild plant foods; and pastoralism.

People from all of the ethnic groups live in extended family homesteads based on patrilocal organization. Among the Bantus, polygyny is common, with 45% of the men over 35 participating in polygynous relationships at any one time. Polygyny is rare among the San speakers. Marriage and reproductive unions, however, are fluid among all the ethnic groups. Multipartnered sexuality is commonplace, and disputes over paternity and child support are common in the tribal court. For all the ethnic groups the norm is for men to marry and become fathers in their 30s.

At the time of the study, there was very little cash economy in the study community. Most men of all ethnic groups over the age of 35 had worked in migratory labor, usually in the mines of South Africa, for an average of 5 years. Many of the ||Anikhwe and Bugakhwe men over the age of 25 had been soldiers in the South African Defence Force during the bush wars of the 1970s and 1980s. Few women, however, had ventured beyond the next community 30 kilometers away. There was no school, clinic, or borehole, and water was drawn from a river source.

The nearest primary school was in the next community, about 30 kilometers away, and children attending that school needed to board there. Although there are no school fees in Botswana, the cost of boarding, uniforms, and books as well as the lost labor made school costly to parents. At any one time, approximately 25% of the children in the community attended primary or secondary school. Those attending secondary school boarded at communities at least 100 kilometers away. Due to the lack of vehicles and roads, children attending school returned home only sporadically.

Historically, the Bantus represented in this community have all had some degree of matrilineality in their social organization with a tradition of the avunculate (Larson, 1970). In particular, this implies that in earlier times a boy's strongest male influence would not be from his father but from his mother's eldest brother. Both ||Anikhwe and Bugakhwe have been strongly influenced by Bantus over at least the last 100 years and also have some degree of matrilineality and the avunculate. The situation is not clear-cut, however, since all ethnic groups in the study community have been under strong political and social influence of Tswana-speaking tribes for at least 200 years. The Tswana have a strong tradition of patrilineality, and their social organization and customs regarding marriage, the family, and child rearing have been codified as Botswana's Customary Law. All disputes are settled using this legal code, regardless of the ethnic origin of the litigants, and this pervasive influence has had a profound impact on the maintenance of social organization and tradition by non-Tswana groups.

Work in this community began in 1992 as part of a dissertation project focusing on the determinants of children's activities. There was an additional field session in this community covering most of 1994, and there have been frequent subsequent visits, with the latest occurring in 2001. A second community that was far more market incorporated, composed mostly of ||Anikhwe and Wayeyi families, was included in the study in 1996 and produced 2 years of data collection. Again, there have been frequent subsequent visits to this community. Future research in these and other communities will focus on the effects of the HIV/AIDS epidemic on the family and child development in Botswana, which has one of the highest HIV prevalence rates in the world (approximately 36% of adults).

Data Collection

The time allocation data used in this chapter consist of instantaneous scan data in combination with data from focal observations of children. The instantaneous scan samples were collected over the course of 11 months in 1992. Extended family homesteads were sampled repeatedly, on a rotating basis, over three 4-hour periods (0600–1000, 1000–1400, and 1400–1800) roughly corresponding to daylight hours. On an hourly time point, the activity, location, and interactants of all residents of the homestead were noted. For residents who were not present, other residents were asked about that person's activity and location, and this information was verified with the focal individual either upon his or her return, or later. Focal observations of individuals away from homesteads were conducted so that each of the major activities in which children engage (as determined by the household samples) were represented adequately. These observations were designed to determine the amount of time people spend in activities when they were away from homesteads, intensity of activity, and return rates for activities.

During observations, the beginning and ending times of all activities were recorded, and instantaneous records of behavior, recorded every 10 minutes, included the identity of coparticipants. In addition, food acquired was weighed, and all food consumed was recorded in terms of species, acquirer, field or herd of origin, and immediate giver. These data consist of repeated observations of 40 boys and 35 girls from 3 to 18 years old. A total of 1,435 observations at 10-minute intervals was made between April and October 1992, for an average of about 3 hours per individual. There was no significant difference in the average for boys or girls. This finding compares with an average of 35 hours of instantaneous scans for each individual (n = 4,184 total observations; n = 2,137 observations on boys; n = 1,947 observations on girls). These datasets were concatenated, and the observations from the instantaneous scans were then weighted to reflect the difference in time scale of the observations.

Data Analysis

No a priori definition of possible play types existed before the initial data collection. As data collection proceeded, a list was created from observed events and continually expanded. Activities coded as various forms of play during data collection were recoded into broader categories. The categories chosen correspond to those used in play research by psychologists: object play, fantasy play, and R&T play. In addition, play was further subcoded into readily identifiable subcategories; these were play pounding, ball games, and two games that we observed children playing with some frequency, called the "aim game" and the "cow game."

Play pounding, a very common activity for girls, is a fantasy imitation of the pounding phase of grain processing (for detailed description, see Bock, 1998, 2002b). Although occasionally played with a real mortar, it is usually played in the dirt with an imaginary mortar and a reed or stick for the pestle. Ball games usually have flexible rather than rigid a priori rules. In the "aim game," which is played by boys, one boy picks a spot on the ground and throws a stick at it. Other boys follow suit. The "cow game" is a fantasy, role-playing game involving cattle. Some children pretend to be oxen and are "yoked" to a "sledge" by other children. Usually the "yoke" is indicated by the use of some leftover twine that is easy to find lying around the yard. The "sledge" can be either a toy sledge, which is often made for children by their fathers, or some other object imagined to be a sledge. The "sledge" is loaded with dirt or other materials, often imagined to be something else, and transported by the "oxen." One child acts as the driver or handler. Occasionally the "oxen" bolt or lose the load or rear up, in imitation of real oxen. The driver must then use the appropriate techniques to calm down the "oxen" and once again get his or her load underway. The data regarding these activities are very reliable, because they are all distinctive and readily identifiable.

The proportion of the total observations of play for each type of play was calculated for each individual. These proportions were then converted to mean minutes per day spent in that activity as (the proportion of time spent in the specific play activity multiplied by the proportion of total time spent in play multiplied by 720 minutes/day). Regression analysis was conducted using SAS version 6.08.

RESULTS

Age and Gender Effects

The first analyses describe the overall age- and gender-specific patterning of time allocation to play. The numbers of observations were aggregated over age groups, and then proportions of total time spent in play and proportions in each type of play were calculated. Table 10.1 shows these results for boys, and Table 10.2 for girls. The proportion of total time spent playing shows a general decline over age. For boys, 4- to 6-year-olds were observed in play 25.7% of

TABLE 10.1. Proportion of Total Time Observed in Play, and Proportion of This for Each Type of Play, for Boys

Age category (years)	All play	Fantasy	Object	Rough and tumble	Play pound	Ball	Aim game	Cow game	Other
0–3	0.195	0.022	0.112	0.090	0.000	0.000	0.000	0.022	0.739
4–6	0.257	0.031	0.169	0.125	0.026	0.024	0.000	0.020	0.571
7–9	0.150	0.089	0.037	0.133	0.044	0.015	0.011	0.022	0.567
10–12	0.126	0.083	0.128	0.101	0.000	0.064	0.043	0.028	0.553
13–15	0.033	0.000	0.000	0.200	0.000	0.000	0.032	0.000	0.768
16–18	0.000	0.000	0.000	0.000	0.000	0.000	0.000	0.000	0.000

TABLE 10.2. Proportion of Total Time Observed in Play, and Proportion of This for Each Type of Play, for Girls

Age category (years)	All play	Fantasy	Object	Rough and tumble	Play pound	Ball	Aim game	Cow game	Other
0–3	0.190	0.065	0.042	0.107	0.054	0.000	0.000	0.018	0.714
4–6	0.319	0.023	0.085	0.111	0.162	0.003	0.000	0.017	0.581
7–9	0.083	0.087	0.116	0.043	0.174	0.000	0.000	0.000	0.536
10–12	0.095	0.063	0.156	0.125	0.219	0.000	0.000	0.031	0.406
13–15	0.016	0.000	0.000	0.000	0.000	0.000	0.000	0.000	1.000
16–18	0.000	0.000	0.000	0.000	0.000	0.000	0.000	0.000	0.000

the time, falling to 3.3% for 13- to 15-year-olds. For girls, 4- to 6-year-olds were observed playing 31.9% of the time, falling to 1.6% for 13- to 15-year-olds. The data for 16- to 18-year-old boys and girls are problematic because of the small number of observations due to school attendance. In boys, fantasy play seems to increase through time, whereas object and R&T play seem to be fairly constant. In girls, object play shows an increasing trend, whereas fantasy play shows no change, and R&T play decreases through time.

For boys, participation in ball games peaks at 10–12 years. As stated above, the observations for ball games are biased by the periodicity of ball availability, and they may not reflect the time allocated to ball games if balls were constantly available. The aim game peaks in the 10- to 13-year-old range, whereas the cow game is fairly constant across age classes. For girls, play pounding is the specific form of play most often observed, and young girls spend a good deal of time in that activity; the cow game is an occasional play activity of young girls.

Lastly, for both boys and girls the category of "other forms of play" is large. These play activities were indeterminately coded. This category includes activities coded as "solitary play" and social play that could not clearly be identified as pretend. For girls, these activities encompass from 40.6% of play activities for 10- to 13-year-olds to 100% for 13- to 15-year-olds. For boys, these activities encompass from 57.1% of play activities for 4- to 6-year-olds to 76.8% for 13- to 15-year-olds.

Figure 10.1 shows the age-specific time allocation to all types of play for boys ages 3–18 years. Each point represents the mean daily time allocation to all types of play for each boy. A first-order OLS regression shows that there is a significant negative trend with age ($n = 40$, $p < .0001$, $R^2 = .366$). Figure 10.2 shows the same relationship for girls ages 3–18 years ($n =$

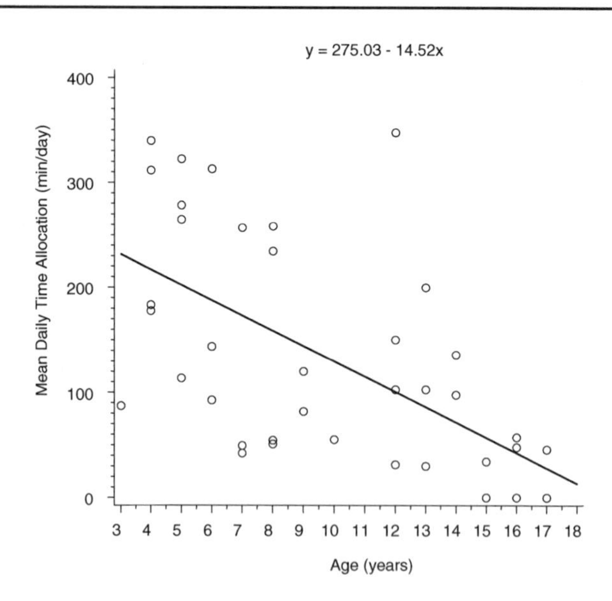

FIGURE 10.1. Age-specific time allocation to all types of play, boys ages 3–18 years.

35, $p < .0001$, $R^2 = .409$). The overall decline over time is consistent with the prediction that there should be an inverse relationship between time allocated to productive and nonproductive activities throughout the developmental process.

Additional analyses were performed to assess the strength of the relationship of age to observations of object, fantasy, and R&T play. There was no age effect on fantasy play for boys ($n = 40$, $p = .137$, $R^2 = .073$) or girls ($n = 35$, $p = .414$, $R^2 = .020$). Boys' object play shows a declining trend over time ($n = 40$, $p = .033$, $R^2 = .210$), as shown in Figure 10.3. For girls, there is no age effect on object play ($n = 35$, $p = .219$, $R^2 = .219$). R&T play showed a decline with age for both boys ($n = 40$, $p = .036$, $R^2 = .110$) and girls ($n = 35$, $p = .005$, $R^2 = .219$), as shown in Figures 10.4 and 10.5.

The aim game and the cow game are primarily boys' activities that are not seen after the age of 12. The cow game has a nonsignificant age effect in this model test, due to the large number of observations of zero ($n = 40$, $p = .361$, $R^2 = .022$). The aim game has a significant effect of age, but also has a large number of zero observations ($n = 40$, $p = .027$, $R^2 = .122$). Play pounding is a girls' activity exclusively, and it is not seen after age 10. There is a very significant age effect, as shown in Figure 10.6 ($n = 35$, $p = .0015$, $R^2 = .282$).

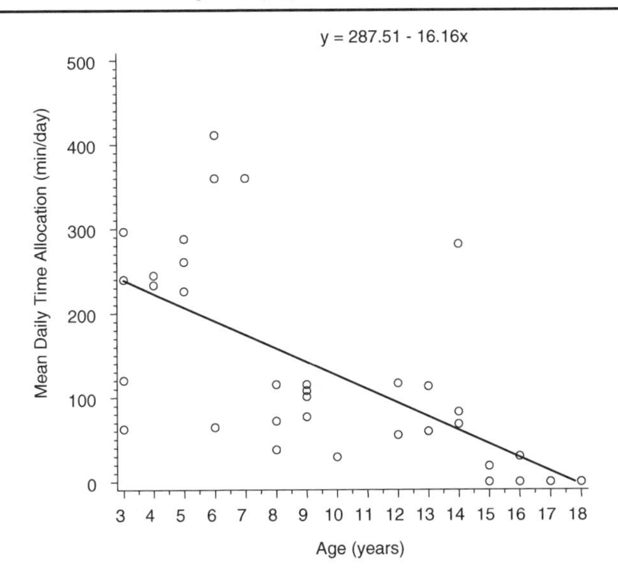

FIGURE 10.2. Age-specific time allocation to all types of play, girls ages 3–18 years.

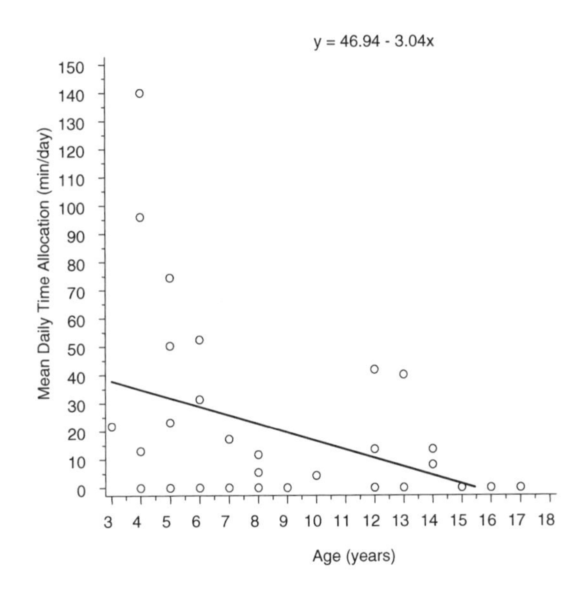

FIGURE 10.3. Age-specific time allocation to object play, boys ages 3–18 years.

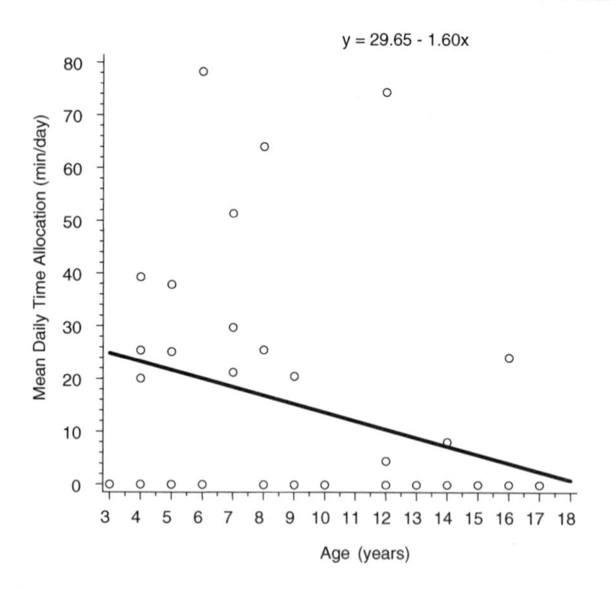

FIGURE 10.4. Age-specific time allocation to rough-and-tumble play, boys ages 3–18 years.

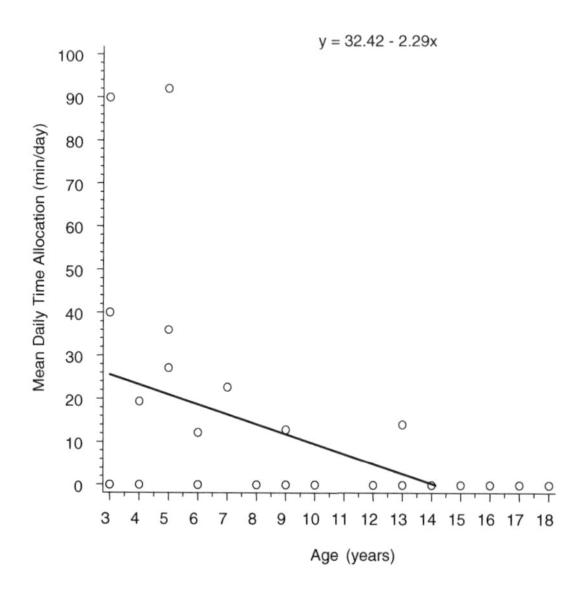

FIGURE 10.5. Age-specific time allocation to rough-and-tumble play, girls ages 3–18 years.

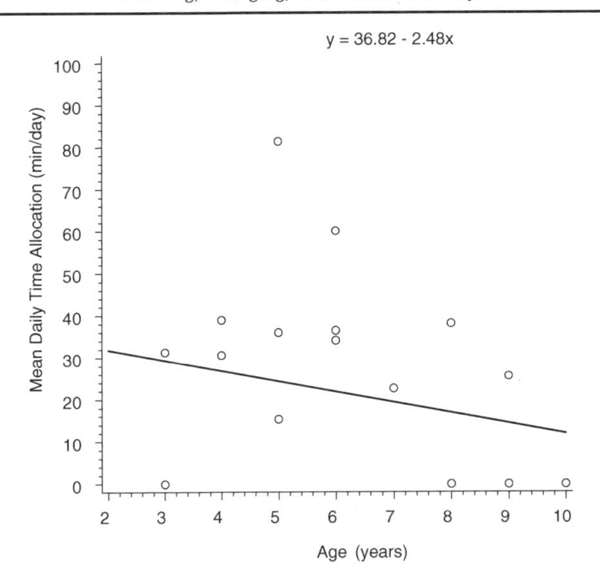

FIGURE 10.6. Age-specific time allocation to play pounding, girls ages 2–10 years.

Work versus Play

As stated above, we should see an inverse relationship between time spent in all nonproductive and productive activities across ages for each gender. The data from the instantaneous scan samples were used to test this prediction. A first-order OLS showed that boys' work is an increasing function of age ($n = 40, p = .079, R^2 = .079$), but this relationship is not statistically significant. Play, however, is a strongly decreasing function of age ($n = 40, p < .0001, R^2 = .410$). There is a crossover point at about 12 years when boys are working more than playing. Girls have a similar but more pronounced relationship (work: $n = 35, p = .007, R^2 = .201$; play: $p = .0006, R^2 = .305$). The crossover point for girls is about 10 years.

In this analysis, work and play are not the only two categories. If we classified activities as productive and nonproductive, for instance, with the total number of observations of nonproductive activities equal to the total number of observations reduced by the number of observations in productive activities, this would not be as interesting a finding. In that instance, there would necessarily be the sort of relationship shown. In this study, the observations included in the play category are only those originally recorded by the data collector as being some form of play. There is no obligatory tradeoff between productive and nonproductive activities as an artifact of data collection. Rather, the above results are clear support for the hypothesis that there is a tradeoff between activities that are immediately productive and those that are related to skill acquisition.

Play Pounding as a Precursor to Pounding

Because play pounding is such a common activity, it is well represented in the instantaneous scan dataset. Children explicitly imitate the motions of grain processing, and the association between the play activity and the real activity is indisputable. If play pounding develops the skills required to be an efficient grain processor, then there should be some tradeoff between the productive activity and the skill acquisition activity. Figure 10.7 shows the results of OLS regression of mean time per day spent play pounding by girls and age ($n = 35, p = .0005, R^2 = .311$), plotted against the results of an OLS regression of pounding and age ($n = 35, p = .0024, R^2 = .261$). There is a crossover at about age 10 that implies that sufficient skills are achieved by that age, such that the opportunity cost to the productive task in skill acquisition is outweighed by the benefits of the child's engagement in the task. I

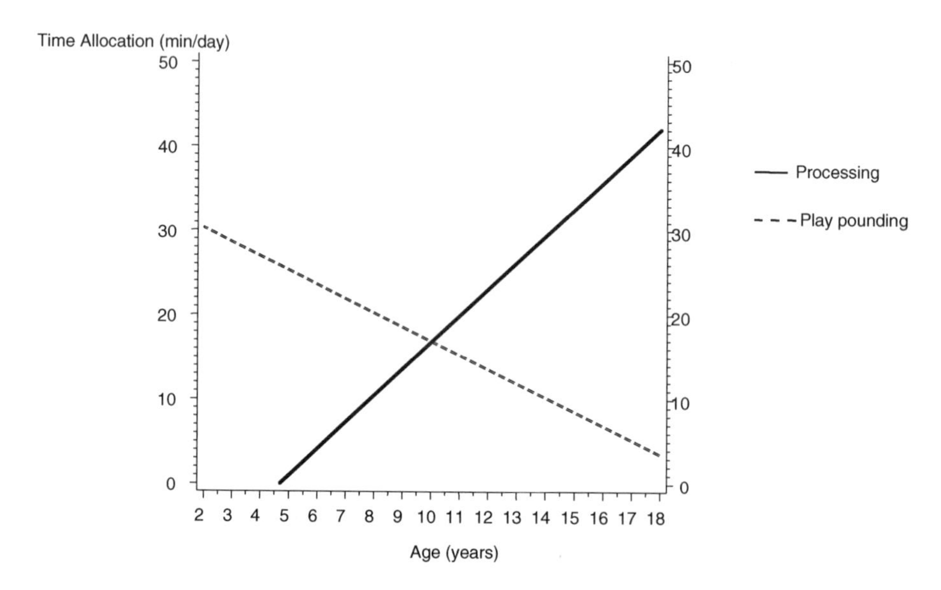

FIGURE 10.7. Age-specific time allocation to play pounding with time allocation to processing grain, girls ages 2–18 years.

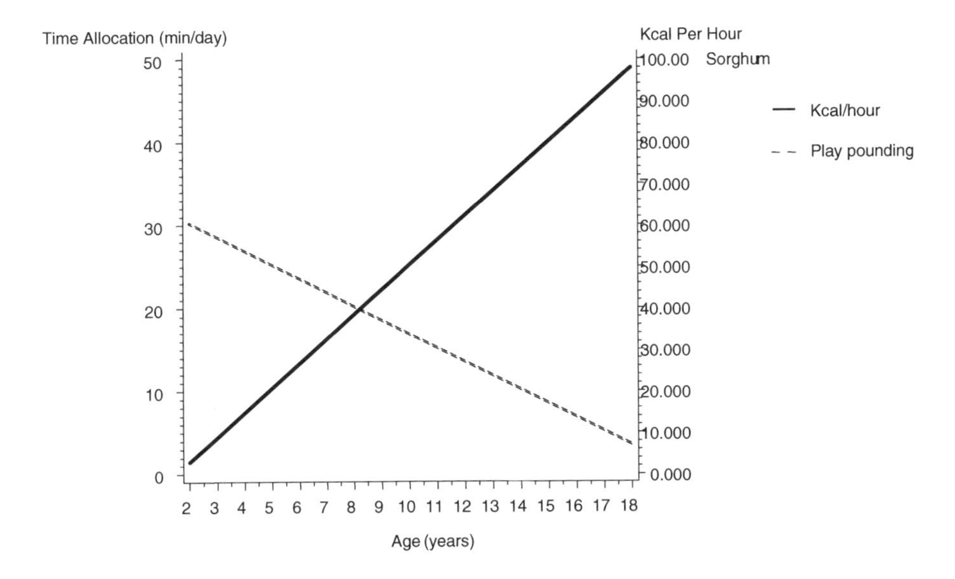

FIGURE 10.8. Age-specific time allocation to play pounding with returns from grain processing, girls ages 2–18 years.

(Bock, 2002b) have demonstrated that girls continue to improve at grain processing as their engagement in performing the task increases.

Figure 10.8 shows time allocation to play pounding in mean daily time allocation, plotted against experimental returns from grain processing for girls. The crossover point is at about 9 years. This finding indicates that the reduction in time allocation to play pounding corresponds directly to improvement in pounding ability. In conjunction with the result shown in Figure 10.7, this finding indicates that girls spend less time play pounding and more time in actual grain processing as they become more proficient at the task (Bock, 2002b).

DISCUSSION

The age- and gender-specific results for object, pretend, and R&T play shed light on the relationship between local ecology and time allocated to play in this community. Girls' time allocation to all forms of play decreases more quickly than boys', with girls playing at very low levels by their early teens. This finding most likely reflects the greater demand for girls' labor in this community (Bock, 2002a, 2002b; Bock & Johnson, 2002a), as occurs in other societies with traditional subsistence patterns (Crognier, Baali, & Hilali, 2001; Hames & Draper, in press; Kramer, 2002; Reynolds, 1991), as well as the earlier maturational trajectory of girls. The frequency of object play, excluding play pounding, shows no consistent patterning by gender in this community. This finding is consistent with studies in other traditional communities (Smith, in press).

When specific forms of object play, such as play pounding, are considered, they show strong gender patterning, with girls spending far more time play pounding, and boys spending far more time in the aim game and playing ball games. One interpretation of these results, arising from the punctuated development model, is that play pounding provides specific preparation for pounding skills; therefore, the direct linkage between these activities, coupled with the relatively short skill acquisition curve to become proficient in pounding (Bock, 2002b), leads to increasing time allocation to play pounding in girls (Bock, 2002a). From this perspective, the aim game might develop skills in the aiming and delivery of projectiles in the context of hunting, and therefore be more prevalent in boys, who are far more likely than girls to hunt or fish with projectile weapons as adults in this society.

Although only among the Bugakhwe in this community is foraging the sole source of food resources, hunting is an important dietary contribution for almost all households. Still, we would expect that the orientation of a household to a specific subsistence regime such as foraging or farming would predict the time allocation to forms of play that develop skills specific to that subsistence regime. My colleague and I (Bock & Johnson, 2004) have shown that, controlling for age and gender, the amount of calories the household derives from farming was a significant negative predictor of the amount of time children spend in the aim game. Children from families with less reliance on farmed products and therefore greater reliance on foraging were significantly more likely to have been observed in the aim game, an activity that may be related to the development of adult competence in hunting. In addition, my colleague and I (Bock & Johnson, 2004) found that children from households with greater reliance on farming for their nutritional requirements were, controlling for age and gender, significantly more likely to be observed in play-pounding activities. Again, this finding is consistent with the hypothesis that children are more likely to engage in play activities that develop skills related to adult competence in tasks commonly undertaken in their subsistence context.

The significant relationships that my colleague and I (Bock & Johnson, 2004) found between subsistence ecology and time allocation to the aim game suggest that there may be additional boys' activities that are preparatory for hunting acumen, because the successful pursuit of prey requires far more contingent skills than projectile delivery (Bock, 2002b; Kaplan et al., 2000; Liebenberg, 1990). There is a limit to the benefits to hunting ability of increased time in practicing projectile delivery. To be an effective predator, individuals must acquire skills in tracking, knowledge of animal behavior, endurance, and bush survival that the aim game does not provide.

The frequencies of R&T play in boys and girls show far fewer differences than found in studies of nontraditional societies (DiPietro, 1981; Geary, 2002; Pellegrini & Smith, 1998). Consistent with findings in other studies, however, is that the highest frequencies of R&T play occur in boys between 7 and 15 years old (Pellegrini & Smith, 1998). In addition to showing higher frequencies of R&T play, boys continue to allocate time to R&T into their teens, whereas girls have largely discontinued this activity by age 13. This discontinuation may reflect the greater demands on girls for labor and the resultant tradeoff favoring immediate productivity. It may also reflect the potential benefits of R&T play, especially play fighting, to boys in becoming effective and successful competitors as men (Geary, 2002; Pellegrini & Smith, 1998; Smith, 1982), in which case the tradeoffs in time allocation favor R&T play for boys but not for girls.

The frequency of pretend or fantasy play, excluding play pounding and the cow game, is not different between boys and girls in this community. For both boys and girls, the frequency of pretend play peaks between 7 and 10 years of age, slightly older than in other published studies (Smith, in press), and reaches very low levels by the teenage years. Play pounding and the cow game are forms of object play that occur within a fantasy context; thus, from an analytical standpoint, these activities have elements in common with both. Smith (1982) argues that different types of play have a training function when learning by doing, or direct practice, is costly in terms of danger or other currencies. Play pounding and the cow game fit this criterion, given that play pounding may be costly in terms of the loss of grain, a valuable commodity, and small children may risk serious injury or death by training on live oxen and cattle. Especially in the cow game, role playing may allow children to experiment with different likely scenario in handling oxen, thereby increasing their acumen with very low cost. In his study of Kpelle children's play in Liberia, Lancy (1996) found that fantasy play developed not only specific skills but also larger-scale social and behavioral attributes through the enactment of scenarios and exploration of contingent solutions.

Because the present study was conducted in a community with a mixed subsistence regime, it is possible to delineate some of the sources of variation in the form of children's play and children's time allocation to play in foraging and agrarian subsistence contexts. In this community, controlling for age, gender, and household demography, children from households with greater reliance on farming spend significantly more time in productive activities (Bock, 2002a). In addition, children are able to perform farming-related tasks at, or near, adult levels far earlier than for foraging-related tasks (Bock, 2002b). This finding suggests that children in subsistence farming contexts will have less time to spend in nonproductive activities such as play, and will have fewer forms of play available that develop adult skills. Taken together, this line of thinking suggests that children in an agrarian economy will play less overall and the frequency of play will decline more rapidly with age than for children in a foraging economy.

My colleague and I (Bock & Johnson, 2004) show that, in fact, in this community as the level of reliance on farmed calories increases, children spend significantly less time playing, controlling for age, gender, school attendance, and family demography. Katz (1991) reported that Sudanese children played less and integrated play with work at higher frequencies as the labor demands of subsistence farming increased with age, a finding similar to that of Punch (2003) on rural Bolivian children in subsistence-farming economies. Although the same set of tradeoffs between immediate benefits of productivity and deferred benefits of skill acquisition are in operation in both foraging and farming economies, these data suggest that the tradeoff strongly favors devoting time to immediate benefits in farming economic contexts and to deferred benefits of skill acquisition in foraging economies (see Bock, 1995, 1998, 1999, 2002b).

As well as providing an additional valuable dataset regarding children's play in a traditional society, the results of this study are consistent with the punctuated development model's predictions that time allocated to children's play will reflect tradeoffs between immediate productivity and skill acquisition across a suite of competing and alternative tasks. The age- and gender-specific time allocation profiles of specific forms of play reflect the costs and benefits of that activity within a given subsistence ecology at a specific point in the life course. This study also demonstrates the value of applying life history models to interpreting not only human universals, but also to understanding phenotypic plasticity in the form of flexible responses to environmental variation.

ACKNOWLEDGMENTS

I would like to thank the Office of the President of Botswana for permission to conduct this research, as well as the people of the study community for so graciously welcoming us into their lives for the last 14 years. Indispensable research assistance was provided by Sara Johnson and Prince K. S. Ndjarakana. Funding was provided by two LSB Leakey Foundation General Grants, the Department of Anthropology and Graduate School of the University of New Mexico, the Andrew W. Mellon Foundation, and the James A. Swan Fund of the Pitt Rivers Museum of the University of Oxford, as well as a National Science Foundation grant to Henry Harpending and Jeffrey Kurland. I am grateful to Sara Johnson and to the editors for their insightful comments and suggestions that greatly improved this chapter.

REFERENCES

Barber, N. (1991). Play and energy regulation in mammals. *Quarterly Review of Biology, 66*, 129–147.

Bekoff, M., & Byers, J. A. (1981). A critical reanalysis of the ontogeny and phylogeny of mammalian social and locomotor play: An ethological hornet's nest. In K. Immelmann, G. W. Barlow, L. Petrinovich, & M. Main (Eds.), *Behavioural development* (pp. 296–336). Cambridge, UK: Cambridge University Press.

Bekoff, M., & Byers, J. A. (1998). *Animal play: Evolutionary, comparative, and ecological perspectives.* Cambridge, UK: Cambridge University Press.

Bird, D., & Bliege Bird, R. (2002). Children on the reef: Slow learning or strategic foraging. *Human Nature, 13*, 269–297.

Bjorklund, D. F., & Pellegrini, A. D. (2002). *The origins of human nature: Evolutionary developmental psychology.* Washington, DC: American Psychological Association Press.

Bliege Bird, R., & Bird, D. (2002). Constraints of knowing or constraints of growing?: Fishing and collecting by the children of Mer. *Human Nature, 13*, 239–267.

Bloch, M., & Pellegrini, A. D. (Eds.). (1989). *The ecological context of children's play.* Norwood, NJ: Ablex.

Blurton Jones, N. G. (1993). The lives of hunter-gatherer children: Effects of parental behavior and parental reproductive strategy. In M.E. Pereira & L.A. Fairbanks (Eds.), *Juvenile primates* (pp. 309–326). New York: Oxford University Press.

Blurton Jones, N. G., Hawkes, K., & O'Connell, J. F. (1999). Some current ideas about the evolution of human life history. In P. C. Lee (Ed.), *Comparative primate socioecology* (pp. 140–166). Cambridge, UK: Cambridge University Press.

Blurton Jones, N. G., & Konner, M. (1973). Sex differences in the behavior of Bushman and London two- to five-year-olds. In R. P. Michael & J. H. Crook

(Eds.), *Comparative ecology and behavior of primates* (pp. 689–750). New York: Academic Press.

Blurton Jones, N. G., & Marlowe, F. (2002). Selection for delayed maturity: Does it take 20 years to learn to hunt and gather? *Human Nature, 13*, 199–238.

Bock, J. (1995). The determinants of variation in children's activities in a southern African community. *Dissertation Abstracts International, 58*(08A), 3188.

Bock, J. (1998). Economic development and cultural change among the Okavango Delta peoples of Botswana. *Botswana Notes and Records, 30*, 27–44.

Bock, J. (1999). Evolutionary approaches to population: Implications for research and policy. *Population and Environment, 21*, 193–222.

Bock, J. (2002a). Evolutionary demography and intrahousehold time allocation: Schooling and children's labor among the Okavango Delta peoples of Botswana. *American Journal of Human Biology, 14*, 206–221.

Bock, J. (2002b). Learning, life history, and productivity: Children's lives in the Okavango Delta of Botswana. *Human Nature, 13*, 161–198.

Bock, J. (2005). What makes a competent adult forager? In B. Hewlett & M. Lamb (Eds.), *Hunter-gatherer childhoods* (pp. 109–128). Hawthorne, NY: Aldine de Gruyter.

Bock, J., & Johnson, S. E. (2002a). Male migration, remittances, and child outcome among the Okavango Delta peoples of Botswana. In C. S. Tamis-LeMonda & N. Cabrera (Eds.), *Handbook of father involvement: Multidisciplinary perspectives* (pp. 308–335). Mahwah, NJ: Erlbaum.

Bock, J., & Johnson, S. E. (2002b). The Okavango Delta peoples of Botswana. In R. K. Hitchcock & A. J. Osborne (Eds.), *Endangered peoples of Africa and the Middle East* (pp. 151–169). New York: Greenwood Press.

Bock, J., & Johnson, S. E. (2004). Subsistence ecology and play among the Okavango Delta peoples of Botswana. *Human Nature, 15*, 63–81.

Bock, J., & Sellen, D. W. (2002). Introduction to special issue on childhood and the evolution of the human life course. *Human Nature, 13*, 153–161.

Bogin, B. (1999). *Patterns of human growth* (Vol. 23). Cambridge, UK: Cambridge University Press.

Boulton, M., & Smith, P. K. (1992). The social nature of play fighting and play chasing: Mechanisms and strategies underlying cooperation and compromise. In J. Barkow, L. Cosmides, & J. Tooby (Eds.), *The adapted mind* (pp. 429–444). Oxford, UK: Oxford University Press.

Briggs, J. (1990). Playwork as a tool in the socialisation of the Inuit child. *Arctic Medical Research, 49*, 34–38.

Byers, J. A., & Walker, C. (1995). Refining the motor training hypothesis for the evolution of play. *American Naturalist, 146*, 25–40.

Charnov, E. L. (1993). *Life history invariants: Some explorations of symmetry in evolutionary ecology.* Oxford, UK: Oxford University Press.

Charnov, E. L., & Berrigan, D. (1993). Why do female primates have such long lifespans and so few babies? *Evolutionary Anthropology, 1*, 191–194.

Cohen, D. (1987). *The development of play.* London: Croom Helm.

Crognier, E., Baali, A., & Hilali, M.-K. (2001). Do "helpers at the nest" increase their parents' reproductive success? *American Journal of Human Biology, 13*, 365–373.

DiPietro, J. A. (1981). Rough and tumble play: A function of gender. *Developmental Psychology, 17*, 50–58.

Eibl-Eibesfeldt, I. (1989). *Human ethology.* New York: Aldine de Gruyter.

Fagen, R. M. (1981). *Animal play behavior.* New York: Oxford University Press.

Fry, D. P. (1990). Play aggression among Zapotec children: Implications for the practice hypothesis. *Aggressive Behavior, 16*, 321–340.

Geary, D. (2002). Sexual selection and human life history. In R. Kail (Ed.), *Advances in child development and behavior* (Vol. 30, pp. 41–101). San Diego, CA: Academic Press.

Gosso, Y., & Otta, E. (2003, June). *Parakanã Indian children at play.* Paper presented at the annual meetings of the Jean Piaget Society, Chicago.

Gould, S. J., & Lewontin, R. C. (1979). The spandrels of San Marco and the Panglossian paradigm: A critique of the adaptationist programme. *Proceedings of the Royal Society of London (B), 205*, 581–598.

Hames, R., & Draper, P. (in press). Women's work, child care, and helpers at the nest in a hunter-gatherer society. *Human Nature.*

Hawkes, K., O'Connell, J. F., Blurton Jones, N. G., Alvarez, H., & Charnov, E. L. (1998). Grandmothering, menopause, and the evolution of human life histories. *Proceedings of the National Academy of Sciences, 95*, 1336–1339.

Hill, K., & Hurtado, A. M. (1996). *Ache life history: The ecology and demography of a foraging people.* New York: Aldine de Gruyter.

Kamei, N. (in press). Play among Baka children in Cameroon. In B. Hewlett & M. Lamb (Eds.), *Hunter-gatherer childhoods.* Hawthorne, NY: Aldine de Gruyter.

Kaplan, H. S. (1996). A theory of fertility and parental investment in traditional and modern human societies. *Yearbook of Physical Anthropology, 39*, 91–135.

Kaplan, H. S., & Bock, J. (2001). Fertility theory: The embodied capital theory of human life history evolution. In N. J. Smelser & P. B. Baltes (Eds.), *The international encyclopedia of the social and behavioral sciences* (pp. 5561–5568). Oxford, UK: Elsevier Science.

Kaplan, H. S., Lancaster, J. B., Bock, J. A., & Johnson, S. E. (1995). Does observed fertility maximize fitness among New Mexican men?: A test of an optimality model and a new theory of parental investment in the embodied capital of offspring. *Human Nature, 6*, 325–360.

Kaplan, H. S., Lancaster, J. B., Hill, K., & Hurtado, A. M. (2000). A theory of human life history evolution: Diet, intelligence, and longevity. *Evolutionary Anthropology, 9*, 156–183.

Katz, C. (1991). Sow what you know: The struggle for social reproduction in rural Sudan. *Annals of the Association of American Geographers, 81*, 48–51.

Konner, M. (1972). Aspects of the developmental ethology of a foraging people. In N. Blurton Jones (Ed.), *Ethological studies of child behaviour* (pp. 285–304). Cambridge, UK: Cambridge University Press.

Kramer, K. (2002). Variability in the duration of juvenile dependence: The benefits of Maya children's work to parents. *Human Nature, 13*, 327–344.

Lancy, D. F. (1996). *Playing on the mother ground: Cultural routines for children's development.* New York: Guilford Press.

Larson, T. J. (1970). The Hambukushu of Ngamiland. *Botswana Notes and Records, 2*, 29–44.

Leigh, S. R. (2001). The evolution of human growth. *Evolutionary Anthropology, 10*, 223–236.

Liebenberg, L. (1990). *The art of tracking: The origin of science.* Cape Town, South Africa: Philip.

Martin, P., & Caro, T. (1985). On the function of play and its role in behavioral development. In J. Rosenblatt, C. Beer, M. Bushnel, & P. Slater (Eds.), *Advances in the study of behavior, 15*, 59–103.

Pellegrini, A. D. (2002). Rough-and-tumble play from childhood through adolescence: Development and possible functions. In P.K. Smith & C. Hart (Eds.), *Blackwell handbook of social development* (pp. 438–453). Oxford, UK: Blackwell.

Pellegrini, A. D., & Bjorklund D. F. (2004). The ontogeny and phylogeny of children's object and fantasy play. *Human Nature, 15*, 23–43.

Pellegrini, A. D., & Smith, P. K. (1998). Physical activity play: The nature and function of a neglected aspect of play. *Child Development, 69*, 577–598.

Pepler, D. J., & Rubin K. (Eds.). (1982). *The play of children: Current theory and research.* Basel, Switzerland: Karger.

Power, T. G. (2000). *Play and exploration in children and animals.* Mahwah, NJ: Erlbaum.

Punch, S. (2003). Childhoods in the majority world: Miniature adults or tribal children? *Sociology, 37*, 277–296.

Reynolds, P. (1991). *Dance civet cat: Child labour in the Zambezi valley.* London: ZED (with Ohio University Press and Baobab Publications).

Roopnarine, J. L., Johnson, J. E., & Hooper, F. H. (Eds.). (1994). *Children's play in diverse cultures.* Albany: State University of New York Press.

Rubin, K. H. (1982). Early play theories revisited: Contributions to contemporary research and theory. In D. J. Pepler & K. H. Rubin (Eds.), *The play of children: Current theory and research* (pp. 4–14). Basel, Switzerland: Karger.

Rubin, K. H., Fein, G., & Vandenberg, B. (1983). Play. In E. M. Hetherington (Ed.), *Handbook of child psychology: Socialization, personality, and social development* (Vol. 4, pp. 693–774). New York: Wiley.

Schwartzman, H. (1976). The anthropological study of children's play. *Annual Review of Anthropology, 5*, 289–328.

Schwartzman, H. (1978). *Transformations: The anthropology of children's play.* New York: Plenum Press.

Slaughter, D. T., & Dombrowski, J. (1989). Cultural continuities and discontinuities: Impact on social and pretend play. In M. N. Bloch & A. D. Pellegrini (Eds.), *The ecological context of children's play* (pp. 282–310). Norwood, NJ: Ablex.

Smith, P. K. (1982). Does play matter?: Functional and evolutionary aspects of animal and human play. *Behavioral and Brain Sciences, 5*, 139–184.

Smith, P. K. (in press). Play types and functions in human development. In B. J. Ellis

& D. F. Bjorklund (Eds.), *Origins of the social mind: Evolutionary psychology and child development*. New York: Guilford Press.

Smith, P. K., & Simon, T. (1984). Object play, problem-solving and creativity in children. In P. K. Smith (Ed.), *Play in animals and humans* (pp. 199–216). Oxford, UK: Basil Blackwell.

Sutton-Smith, B. (1971). The role of play in cognitive development. In R. Herron & B. Sutton-Smith (Eds.), *Child's play* (pp. 252–260). New York: Wiley.

Trivers, R. (1972). Parental investment and sexual selection. *Nature, 112,* 164–190.

Walker, R., Hill, K., Kaplan, H., & McMillan, G. (2002). Age dependency of strength, skill, and hunting ability among the Ache of Paraguay. *Journal of Human Evolution, 42,* 639–657.

Worthman, C. M. (2000). Evolutionary perspectives on the onset of puberty. In W. R. Trevathan, J. J. McKenna, & E. O. Smith (Eds.), *Evolutionary medicine* (pp. 135–163). New York: Oxford University Press.

PART VI

CONCLUSION

CHAPTER 11

Play in Great Apes and Humans

Reflections on Continuities and Discontinuities

PETER K. SMITH AND ANTHONY D. PELLEGRINI

This book has brought together overviews of different forms of play in great apes and in various types of human societies. By focusing on great apes, we have tried to avoid too broad a panorama of animal play, instead concentrating on those species that are phylogenetically closest to humans. By deliberately drawing on data from various types of human societies, we have tried to avoid too narrow a preoccupation with data from modern industrial societies. Although the vast majority of our human data comes from the latter, we believe it vital to examine play in more traditional hunter-gatherer, foraging, and tribal societies, if we are to be in a position to address questions on the evolution of play and on its functional significance in apes and humans.

We have also focused mainly on three types of play. Play fighting, or rough-and-tumble play (R&T), is a predominant form of social play in apes and common (if relatively neglected for study) in human children. Object play is found in apes and is common in human children. Pretend or fantasy play is rarely seen in apes but is almost a paradigmatic form of play in young children. These three types form a spectrum on the ape–human comparison and are an obvious focus for reflections on continuities and discontinuities. (Of course, there are other forms of play: for example, exercise play, sexual play, social contingency play, and in humans at least, language play.) Of these three types, play fighting could be argued to have clear functions in apes, perhaps less clear function in humans; object play could be argued to show some continuity of function; pretend play, in contrast, is of doubtful function in apes (given that it is so rare), whereas its function in human children is often assumed but remains uncertain in nature and extent.

In this final chapter we first provide an overview of the contributions, then offer reflections on the issues raised by the material and on the extent of continuity and discontinuity in great ape and human play.

A REVIEW OF THE CONTRIBUTIONS

An important theoretical backdrop for this book is provided by *Patrick Bateson*, who has done seminal work on play (primarily of cats) and, more generally, on the nature of development. Most instructive for developmental psychologists, we think, is his distinction between immediate and delayed benefits associated with play and of the impact of behavior on evolutionary functions. Recently, he has turned his attention to the ways in which evolutionary theory can be used to inform the study of human development. In this work, he has been very careful to avoid "main-effect models" for the influence of genes on the behavior of humans, or indeed, any other animals.

Bateson argues that play can be reliably measured and that the different forms of human play probably have different functions, a conclusion supported by developmental evidence from other animals. He also argues that the roles of play in evolution raise quite different issues. The active role of behavior in evolution is well recognized in contexts ranging from mate choice to niche construction; increasingly, the importance of organic selection is seen as being very important. The organic selection process involves ontogenetic adaptability, followed by Darwinian evolution of behavior patterns that generate the same result at lower cost. Organic selection provides the most likely route by which play could have driven the evolution of behavior in the great apes and humans. Bateson suggests that those aspects of play that are creative or break out of local optima are especially promising candidates for driving evolution. Bateson's argument is that aspects of play can, indeed, increase the total sum of spontaneously emitted behavior patterns that serve to solve complex problems.

Kerrie Lewis reviews the literature on the social play of both captive and wild great apes. These primates have been reported to engage in almost all the nonhuman primate play behavioral repertoire, including R&T, play mothering, pseudosexual play, and adult play interactions.

Lewis shows that all species devote significant proportions of their daily time budget to play behaviors. However, there are distinct age patterns, with peaks in social play in infancy and the juvenile period before weaning; and sex differences, with social play of older male juveniles being rougher and more aggressive.

Lewis presents recent research showing that social play coevolved with certain aspects of brain function, suggesting a link between both cognitive and motor aspects of playful interactions. Apes are high in social-play frequencies, and they also have a relatively large neocortex, which has in turn been linked to group size and social complexity. The social play of great apes may offer significant insights into the lives of the animals that perform it, in terms of capacity for innovation, social affiliations, dominance relationships, cultural transfer, and cognitive capacities. Social play is argued to enter into the transfer of cultural skills, such as tool use, and it may be involved in the determination of dominance rank and social relationships, including possible paternity. In addition, the great apes' use of various gestures and facial expressions in their social play interactions may offer insight into the evolution of smiling, laughter, and arguably, even humor.

Douglas Fry, an anthropologist by training, overviews one aspect of social play, R&T, in human juveniles. Certain common elements of human R&T are apparent in widely distributed cultures (when information exists): wrestling and grappling, restrained hitting and punching, and chasing and fleeing. These behaviors usually occur in conjunction with laughs, smiles, play faces, and at times, playful vocalizations or speech indicating the fantasy elements of play. These forms are strikingly similar to those noted by Lewis for the great apes. The human cross-cultural data are limited both in number of relevant sources and in the detail of reporting, but the available evidence reflects these common features.

Boys tend to engage in more R&T—especially in the roughest forms, such as wrestling—than do girls. Although data are not available for many cultures, the evidence suggests that a tendency among children, especially boys, to engage in R&T is probably a human universal. At the same time, cultural variations on the R&T theme exist. Fantasy elements reflect cultural features: for example, British children shoot imaginary guns, Zapotec (Mexico) children play at rodeo by attempting to subdue a "bull" (played by another child), and Santal children play the *sim sim* game of "chickens" versus a "kite." In relatively nonviolent societies, such as the Semai and La Paz Zapotec, adults discourage children from engaging in R&T. The milder forms of R&T among the Semai and relatively less R&T among La Paz Zapotec in comparison to San Andrés Zapotec illustrate that R&T is open to various social influences.

Research on R&T in children and adolescents suggests that both practice and dominance functions are likely, perhaps primarily in that developmental sequence. For humans, the practice of fighting skills may be more complex than in other species. Humans may have to practice not only fighting maneuvers but also restraint within a social world that includes rules for fighting and dispute resolution. It is interesting to contemplate how much aggression among humans is restrained, curtailed, or limited through social conventions, enforced by the participants themselves and by other members of the group. Contests are one obvious example of restrained aggression and may be similar to the R&T in adolescent boys: Both provide a way to resolve dominance struggles, with minimal risk to the participants. At the same time, contests are by no means a universal solution to this problem; humans generally, and nomadic foragers in particular, use many ways to handle conflict, some more competitive than others. As suggested by the data on the La Paz Zapotec and the Semai, and reinforced by the existence of highly peaceful nomadic hunter-gatherer societies such as the Batek and the Paliyan, the expression of both aggression and R&T can be reduced to very low levels. At the same time, the manifestation of the R&T theme in children around the world suggests the presence of evolved functions for the behavioral pattern, of which the practice of fighting skills, broadly conceived, and dominance assertion seem likely candidates.

Object play of nonhuman primates has a long and interesting history. Most of us have images of Jane Goodall's chimpanzees fishing for termites. *Jacklyn Ramsey* and *William McGrew* review studies of object use and play among captive and wild apes. They note that previous studies of object play, both in nature and captivity, have failed to separate play from the broader category of object manipulation. The current published literature on the subject in great apes shows that such distinctions are often ignored, or if acknowledged, are not addressed.

Ramsey and McGrew tackle this omission in this chapter, as they present the results from two studies of object play in great apes. The results from the first study, of five infant chimpanzees in the wild, show a reassuring congruence between object play by youngsters and the elementary technology characteristic of this and most other chimpanzee populations. In the second study, they compared chimpanzees and gorillas in captive settings. Gorillas show much less object play in the wild, and this species difference was replicated under artificial conditions with novel and familiar objects present. Most of the captive-born apes had never seen an African forest and so could not be aware of the species difference between *Pan* and *Gorilla* in the natural occurrence of elementary technology. Nor had the wild-born apes been given the chance to fish for termites, sponge for water, crack nuts, and so on, in captivity, and so act as models for their captive-born counterparts. Thus, the differences found are impressive for their consistency in surroundings that can be considered both enriched (by captive standards) but impoverished (by natural standards).

Ramsey and McGrew also point out ethical reasons for studying object play in apes: We need to identify norms from nature in order to provide suitable environmental enrichment for their captive counterparts. If wild apes have material culture, and if we must keep apes in captivity, then we should avoid impoverishing their environment more than is necessary.

Anthony Pellegrini and *Kathy Gustafson* also differentiate forms of object manipulation and provide time-budget information on these activities for object play of human juveniles. They document the occurrence of play, exploration, construction, and tool use in preschool children's everyday life in their nursery classrooms, across an entire school year. They found that tool use, like play, occurs at a moderate level during children's playtime in nursery school. Their results also reinforced the position that exploration, construction, play, and tool use are separate constructs.

Pellegrini and Gustafson also examined the extent to which the cost associated with different types of object use predicted facility with constructing and using tools to solve convergent problems (lure retrieval tasks) and divergent problems (fluency in generating alternate uses for objects). Unlike previous studies of this kind, criticized for lack of ecological validity (Smith & Simon, 1984), the authors used observational data of constructive play and tool use gathered over the previous year (rather than manipulating play conditions for a very short period of time). They also controlled statistically for spatial intelligence. Their results suggest that time spent by children in construction play may be providing important practice and learning opportunities for the development of problem-solving abilities. Total tool use was also a significant predictor of problem solving, on one task.

Juan-Carlos Gómez and *Beatriz Martín-Andrade* review the existing evidence about the occurrence of fantasy play in nonhuman primates, especially anthropoid apes. They begin with a discussion of how imaginative or fantasy play has been defined by different authors in an attempt to derive objective criteria for the identification of this kind of behavior and the evaluation of claims by students of primates. Then they review the observations of behaviors that have been purported to qualify as fantasy play by different authors studying nonhuman primates, organizing their review into observations of natural behaviors, observations in captive settings, and observations in hand-reared primates and "linguistic" apes (i.e., apes that have been trained in the use of artificial symbols).

Fantasy play appears to be very rare in wild apes. There is a range of examples in captive apes, but interpretation of the behaviors is difficult. We can take a skeptical viewpoint as to whether simple imitative actions, such as cuddling a doll, are really "pretend" in a metarepresentational sense, just as we can wonder about the same issue for human infants. The authors conclude that the most convincing (or, at least, the more human-like) examples of fantasy play come from symbolically trained apes, although they are not necessarily the product of direct training but, rather, a by-product of the training process. On the other hand, it is possible to observe prerequisites or precursors of imaginative play in some behavior of untrained apes. Based on this evidence, the authors propose that human fantasy play may have evolved from a combination of precursor behaviors and abilities related to symbolic and explicit representation, only some of which are shown by modern apes.

The study of play in human juveniles has been dominated by the study of pretend play, perhaps reflecting the influence of Piaget on child and developmental psychology. *Peter Smith* outlines the developmental trajectory of play and then focuses on pretend play, debating its functional importance. He argues that much writing and research has been distorted by a "play ethos" whereby play is regarded as crucial to the healthy development of human children; this claim is often made despite equivocal evidence. He proposes three possible models: (1) no special function of pretend play, (2) a function of facilitation among many other useful experiences, and (3) an essential function, the "play ethos." He suggests that the correlational and experimental evidence lend the most support to the second model, regarding any causative influence from pretend play to other cognitive or social skills, including theory of mind.

However, the correlational and experimental studies are all from modern industrialized societies, where pretend and sociodramatic play are often encouraged and fostered by parents in homes and teachers in nursery schools, to an extent greatly exceeding what children probably experienced in traditional societies. In these traditional societies, pretend play occurs but with much less (if any) adult scaffolding; it usually focuses on actual work roles and activities of older people in that society, rather than on "fantastic" figures from the media, as are often featured in Western children's pretend play. Thus, Smith argues, we cannot generalize any facilitating effects of pretend and sociodramatic play on development found in modern societies to traditional societies. Whatever any original function for pretend play might have been (in an evolutionary sense), in modern societies we appear to be "co-opting" pretend and sociodramatic play as convenient vehicles for facilitating certain skills we deem desirable to develop in children.

The foraging way of life (also known as hunting-gathering) is regarded as the environment of human adaptedness (prevalent for well over 90% of the existence of *Homo sapiens*). Our basic psychological mechanisms have been shaped in this context, and it could be argued that study of foraging communities may provide a unique opportunity to learn about the environmental constraints acting on our ancestors and the behavioral strategies they developed to deal with such constraints. *Yumi Gosso, Emma Otta, Maria de Lima Salum e Morais, Fernando José Leite Ribeiro,* and *Vera Silvia Raad Bussab* present a review of child play among extant foragers, both Africans (!Kung, !Ko) and Brazilian Indians (Xikrin, A'uwe-Xavante, Xocó, Guayaki, Kaingáng, Mehinaku, Tukano, Canela, and Parakanã).

Few human groups still lead an exclusively foraging way of life, and present-day foragers are not exact equivalents of our ancestors. The !Kung live at the edge of the Kalahari desert; in the past, in contrast, they probably lived in a wealthier savannah environment. Likewise the Parakanã Indians conduct small-scale subsistence agriculture, although hunting and gathering are routine activities responsible for an important part of their diet. Nevertheless, these and similar communities are much more representative of our ancestral condition than modern urban societies.

Data about different forms of play (exercise, construction, social contingency, R&T, fantasy, and games with rules) are reviewed in their chapter. There is particularly detailed information on the Parakanã Indians studied by Gosso; Parakanã children exhibit much exercise play, relatively little R&T, and some fantasy play focused on actual adult social and cultural activities. Similarities and differences in these categories among various foraging groups are discussed.

Gosso and colleagues are circumspect about the functions of play in foraging peoples. An interesting observation is that children's play may be useful for adults, by keeping children in their autonomous world (perhaps minded by older siblings) and allowing parents to get on with subsistence activities or infant care unimpeded by their demands. Gosso and colleagues also question whether play is important for learning specific skills that are imitated, rather than for generally but gradually imbibing a "hidden curriculum" of cultural practices and meaning, emotional and social skills.

John Bock presents data on the different forms of play performed by male and female juveniles among a pastoral people in Botswana. He reviews life-history approaches and uses the punctuated development model to explain age- and sex-specific findings for object, pretend, and R&T play in relationship to the local ecology.

Bock's analysis clearly suggests the role of parental interests and cultural pressures on the work–play balance in these children's lives. Time allocated to play in this community seems explicable in terms of the tradeoff between skill acquisition and the ability and need for productive work in the economy. This is especially the case when a specific form of object play, play pounding, is considered in relation to demand for older girls' labor in this community. Bock suggests that play pounding is specific preparation for later pounding skills, tolerated or encouraged until actual pounding becomes more productive than further skill acquisition. Similarly, the "aim game" might develop skills in the aiming and delivery of projectiles in the context of hunting, and thus is more prevalent in boys, who are far more likely than girls to hunt or fish with projectile weapons as adults in this society. Likewise, the potential benefits of R&T play, especially play fighting, for boys may be found in the play's rehearsal of skills that will later produce effective and successful competitors. This study demonstrates the value of applying life-history models to interpreting not only human universals, but also to understanding phenotypic plasticity in the form of flexible responses to environmental variation.

CONTINUITIES AND DISCONTINUITIES

Together, the chapters demonstrate continuities and discontinuities between play in great apes and humans. There are continuities in the primary forms of play: exercise play and R&T, object play, and fantasy or pretend play (to a limited extent). There are consistencies in age trends and sex differences, when considered in an ecological and life-history framework. There are important discontinuities, or at least marked differences in scale, when we come to examine the nature of pretend play, the role of parents and societies in cultural transmission, and how play may serve a function in such cultural transmission.

Exercise play appears as a common play form, even though it has received little research attention (see Pellegrini & Smith, 1998). It is common in the great apes and in human children; Gosso and colleagues (Chapter 9, this volume) comment on its high incidence in the Parakanã Indians. Arguably, exercise play is the original form of mammalian play, functioning in a broad sense to maintain neural development, motor control, and physical fitness and stamina (Byers & Walker, 1995). We believe there is a broad continuity here, from the simple jumping, bounding, and play boxing of kangaroos and wallabies (Watson, 1998) to the locomotor play of apes and human children. In a reasonably safe environment and with adequate nutrition, the benefits of locomotor play exceed the time, energy, and danger costs by providing these benefits for present and later growth and fitness. In humans, moreover, the relatively little interest in locomotor play means that adult sanctions (either to encourage it or discourage it) are at a rather low level.

Play fighting, or R&T more generally, also provides an example of some continuity. Again, this is a common mammalian behavior, albeit with species variations in nature and frequency. More characteristic of males than females, it is commonly seen as either practice for fighting and hunting skills later in adolescence or adulthood or as a part of establishing dominance in competitive social groups.

In great apes, and indeed in other mammalian species, cost–benefit arguments suggest that play fighting probably functions to afford practice in fighting or hunting skills and perhaps in dominance assertion. Lewis (Chapter 3, this volume; see also Lewis & Barton, 2004) sees social play as a forum for practicing, modifying, and maintaining such skills, and she links this to the growth of the cerebellum and neocortex. The evidence for a dominance function in human juveniles comes from observations of the nature of R&T and developmental changes with age (Fry, Chapter 4, this volume), the nature of partner choices (Humphreys & Smith, 1987), and possible evidence of "cheating" in R&T (Fagen, 1981; Neill, 1976). Fry mounts a compelling argument that, in human foraging societies, adolescent "contests" are somewhat intermediate in form between play fights and real fights, and function directly in establishing status for males in adolescent groups (clearly, with physical strength being an important advantage). The extent of this intermediate form varies greatly with culture, as Fry documents from his own research.

However, Gosso and colleagues (Chapter 9, this volume) argue that

> play fighting, so important in so many mammals, may have lost nearly all of its relevance for humans. . . . In our view, at some early point of human evolution, those benefits resulting from superior fighting abilities must have declined. Vertical hierarchy gave way to horizontal cooperation among males (and among females). Access to females ceased to be the direct or indirect result of fights. A unique food-sharing pattern, devoid of priorities and disputes, made obsolete the advantages of superior fighting abilities. (p. 233)

More specifically, they believe that "intragroup real fighting is not a human pattern. Intergroup aggression is a very different thing, and it is a human pattern. However, it is done by means of weapons, and we think it has been so for a very long time" (Y. Gosso, personal communication, May 28, 2004). On this argument, physical strength and speed have little relevance when intertribal warfare includes weapons, and within a tribal group, an individual physically aggressive to others would be shunned. Thus, any function of play fighting may be "vestigial."

These arguments are interesting, but we believe them to express a culturally specific view. First, we would note that strength and speed remain useful attributes or skills, even for weapon use, and especially for technologically simple weapons such as spears, knives, and bows. It takes a lot of strength to pull a bow to maximum capacity! Second, regarding within-group aggression, it is indeed the case that in some human societies, such as the Semai and the Parakanã, levels of both real fighting and play fighting are low; this may arguably be relatively truer of hunter-gatherer societies than of settled communities. But play fighting has persisted as a universal human trait; ethnographic accounts, such as that of Chagnon (1968) among the Yanamamo Indians in Brazil, show the continuing importance of fighting ability not only for intergroup conflict, but also for social status and consequent reproductive success in males. We therefore view R&T as a facultative adaptation that can be adjusted to the level of payoff for (usually male) fighting ability and dominance assertion in adolescents and adults in the population.

Object play is much rarer in mammalian species but relatively more common in the primates and great apes. There are problems in distinguishing it from exploration and tool use, and more information is needed, as Ramsey and McGrew, and Pellegrini and Gustafson, document. The sex differences in object play, as well as the species differences, do seem to relate to the kind of technology used by adults in the community. It is thus tempting to see object play as specifically functional in these respects; that is, it may "serve as practice as juveniles learn complicated hunting and gathering skills" (Pellegrini & Bjorklund, 2004, p. 26).

Although plausible, is there good evidence for this view? After all, given the clever imitative abilities of juvenile apes and humans, it is hardly surprising that their object play imitates adult types of tool use. The distinctive sex and species differences do support the functional argument, but in addition, Pellegrini and Gustafson present empirical evidence: Children who engage in a lot of construction play in nursery school also perform well in problem-solving tasks requiring tool use (even with spatial ability controlled for). Their findings suggest quite specific links between play forms and tool-using tasks. Furthermore, at a broader cultural level, Bock finds tradeoffs between play and work activities that are consistent with the view that the play pounding of younger girls provides some skill benefit and is tolerated by adults until actual work productivity would outweigh this future play benefit.

Pretend play provides the most discontinuity between great apes and humans; at least in its full metarepresentational form, it depends on symbolic capacities that are attained by all normally developing human children, but at which great apes, even when enculturated or linguistically trained, are only at the threshold. Having said that, there are certainly parallels between the simple examples seen in apes (Gómez & Martín-Andrade, Chapter 7, this volume), and those in children ages 15 months to 2 years; both usually involve simple imitative actions of eating, drinking, cuddling a "baby," and so forth. In both apes and humans, there is no need to posit a full symbolic capacity for these actions, although clearly they are a steppingstone to full symbolic pretend play.

As an example, consider the cuddling of a doll or baby. In many primates, allo-mothering is common; a juvenile grabs an infant and carries it around, hopefully cuddling it (sometimes, not treating it so well; Quiatt, 1979). At least part of the functional significance of this behavior may be the practice benefits the juvenile obtains (other benefits may include relieving the mother of the task for a while, if the allo-mothering is benign—often by a relative; or harm to a potential competitor, if the allo-mothering is abusive—generally by a female from another matriline). A younger primate imitating this behavior through "maternal play" is thus copying a common behavior of older peers.

These simple forms of fantasy play often overlap with object play, as in Bock's (Chapter 10, this volume) descriptions of "play pounding." They are imitative of adult activities and arguably have some practice benefits, as described for object play.

However, human pretend and sociodramatic play goes well beyond these simple forms, as children ages 3–4 years demonstrate. A joy to ethnographic and developmental researchers, preschool children show great ingenuity and some creativity in their play narratives, especially in Western societies where such play is often encouraged and where there is stimulation from fantasy themes on TV, film, and video, and from specially adapted toys.

It is tempting to view such play as fundamentally important in development—the "play ethos" found in 20th century Western societies (see Smith, Chapter 8, this volume). But we need to bear in mind other possibilities. Bock (Chapter 10, this volume) distinguishes functionalist and neutralist ideas of play, and Gosso and colleagues (Chapter 9, this volume) point out the significant differences between modern urban societies and those of foraging communities. Gaskins (1999, pp. 49–50), commenting on pretend play in Mayan Indian children, points out that it is "less frequent and less socially dominant" than for American children, but that "one should conclude not that there is therefore a deficit in Mayan children but that they spend their time on other, more culturally significant activities." Indeed, in the Mayan context, engagement in a lot of pretend play might have negative consequences, such as "erosion of intrinsic motivation for real-world tasks" (p. 58).

All these contributors point to the importance of adults (often, parents) in their attitudes toward children's play. Whether a child engages in play—perhaps pretend play, especially—varies greatly by culture and by the extent to which adults discourage such play (by imposing work or caregiving demands on even young children), tolerate it for a while (as a way of reducing direct child demands on them for caregiving), or actively encourage it (as a way of developing cognitive and social skills, a form of "parental investment").

Adult or parental influence on play is not specifically human. Breuggeman (1978) demonstrated how rhesus monkey mothers sometimes used play to distract youngsters from unwanted weaning. But generally, nonhuman primate "manipulation" of play for other ends seems relatively slight. It also appears to be rather infrequent in hunter-gatherers (Gosso et al., Chapter 9, this volume), but it would seem that the parents in the tribes studied by Bock (Chapter 10, this volume) were, to some extent, calibrating their encouragement or discouragement of female play pounding according to individual development and economic need (and discouraging play more in societies where farming placed a greater premium on child labor).

Parental manipulation of play has certainly reached a peak in modern industrial societies. Not only does the play ethos influence many parents and many educators of preschool children, in addition, the toy industry has its own agenda in encouraging consumption of toys and games, sometimes with supposedly educational value. It thus becomes difficult to dissociate the possible benefits of play to the individual child—in some abstract individualized sense that nevertheless retains some meaning (e.g., physical prowess, tool-using skill, theory-of-mind ability)—from the child's role as a social being with parents and a community. A child's relationships with parents and community are likely to be as important as any individualized skill acquisition.

We do not share identical views on the issue of the functional nature of pretend play. One of us (A.P.), having documented time and energy costs of pretend play in children, believes that "fantasy play, because of its social nature, has an immediate function of helping juveniles take the perspective of other players" and "[it] does indeed serve an immediate function. Given this level of cost, it would not have been naturally selected for if it did not" (Pellegrini & Bjorklund, 2004, pp. 29, 37). The other (P.K.S.), while acknowledging the force of an evolutionary argument, remains skeptical about any strong function of pretend play, once related benefits of object play or exercise and R&T play (which often have fantasy components) are factored out. He remains more impressed by the cultural variations in pretend play and contends that the specificity with which it is viewed by adults means that it may or may not facilitate acquisition of useful skills in different cultural contexts.

Interweaving these debates are continuing issues about the more precise nature of any play-derived benefits. There are differing views on the neutralist or functionalist position, and on whether a functionalist position implies a necessary function or simply a facilitative (nonessential) one. And, are any functions specific to the particular actions or skills practiced or imitated, or do they reflect general learning about social and cultural requirements? There are differing views on this issue in this book. Is the truth somewhere between the two views?

Bateson (Chapter 2, this volume) points out the complexities of individual–environment and species–environment interaction in determining evolutionary change, even within the confines of genetic evolution. His analysis embraces learning, the ease of acquisition of learned skills, and plasticity in development. Nevertheless, the kinds of cultural influence and transmission seen in humans, and amplified in modern societies, also implicate nongenetic means of transmission. This book may provide a rough sketch of the area, in seeking to understand the evolution and functions of play in great apes and humans. Further empirical data in many areas will help accomplish this endeavor, as will further theoretical advances that integrate genetic and cultural processes of evolution into a larger framework for understanding human behavior and its historical and cultural variations.

REFERENCES

Breuggeman, J. A. (1978). The function of adult play in free-ranging Macaca mulatta. In E.O. Smith (Ed.), *Social play in primates* (pp. 169–191). New York: Academic Press.

Byers, J. A., & Walker, C. (1995). Refining the motor training hypothesis for the evolution of play. *American Naturalist, 146*, 25–40.

Chagnon, N. (1968). *Yanamamo: The fierce people.* New York: Holt, Rinehart & Winston.

Fagen, R. M. (1981). *Animal play behavior.* New York: Oxford University Press.

Gaskins, S. (1999). Children's lives in a Mayan village: A case of culturally constructed roles and activities. In A. Göncü (Ed.), *Children's engagement in the world: Sociocultural perspectives* (pp. 25–61). New York: Cambridge University Press.

Humphreys, A. P., & Smith, P. K. (1987). Rough-and-tumble, friendship, and dominance in schoolchildren: Evidence for continuity and change with age. *Child Development, 58*, 201–212.

Lewis, K. P., & Barton, R. A. (2004). Playing for keeps: Evolutionary relationships between the cerebellum and social play behaviour in non-human primates. *Human Nature, 15*, 5–22.

Neill, S. R. St. J. (1976). Aggressive and non-aggressive fighting in twelve-to-thirteen year old preadolescent boys. *Journal of Child Psychology and Psychiatry, 17*, 213–220.

Pellegrini, A. D., & Bjorklund, D. F. (2004). The ontogeny and phylogeny of children's object and fantasy play. *Human Nature, 15*, 23–43.

Pellegrini, A. D., & Smith, P. K. (1998). The development of play during childhood: Forms and possible functions. *Child Psychology and Psychiatry Review, 3*, 51–57.

Quiatt, D. (1979). Aunts and mothers: Adoptive implications of allomaternal behavior of nonhuman primates. *American Anthropologist, 81*, 310–319.

Smith, P. K., & Simon, T. (1984). Object play, problem-solving and creativity in children. In P. K. Smith (Ed.), *Play in animals and humans* (pp. 199–216). New York: Blackwell.

Watson, D. M. (1998). Kangaroos at play: Play behavior in the Macropodoidea. In M. Bekoff & J. A. Byers (Eds.), *Animal play: Evolutionary, comparative, and ecological perspectives* (pp. 61–95). Cambridge, UK: Cambridge University Press.

Index

Page numbers followed by an *f* indicate figure; *t* indicate table

G

H

I

J

K

L

M

N

O

R

S

T